Physics in Motion

STEM Road Map
for Elementary School

Grade
K

Physics in Motion

Grade K

STEM Road Map
for Elementary School

Edited by Carla C. Johnson, Janet B. Walton, and
Erin Peters-Burton

National Science Teaching Association

Arlington, Virginia

National Science Teaching Association

Claire Reinburg, Director
Rachel Ledbetter, Managing Editor
Andrea Silen, Associate Editor
Jennifer Thompson, Associate Editor
Donna Yudkin, Book Acquisitions Manager

ART AND DESIGN
Will Thomas Jr., Director, cover and
    interior design
Himabindu Bichali, Graphic Designer, interior
    design

PRINTING AND PRODUCTION
Catherine Lorrain, Director

NATIONAL SCIENCE TEACHING ASSOCIATION
David L. Evans, Executive Director

1840 Wilson Blvd., Arlington, VA 22201
www.nsta.org/store
For customer service inquiries, please call 800-277-5300.

NSTA is committed to publishing material that promotes the best in inquiry-based science education. However, conditions of actual use may vary, and the safety procedures and practices described in this book are intended to serve only as a guide. Additional precautionary measures may be required. NSTA and the authors do not warrant or represent that the procedures and practices in this book meet any safety code or standard of federal, state, or local regulations. NSTA and the authors disclaim any liability for personal injury or damage to property arising out of or relating to the use of this book, including any of the recommendations, instructions, or materials contained therein.

Library of Congress Cataloging-in-Publication Data
Names: Johnson, Carla C., 1969- editor. | Walton, Janet B., 1968- editor. | Peters-Burton, Erin E., editor. |
    National Science Teaching Association, issuing body.
Title: Physics in motion, grade K : STEM road map for elementary school / edited by Carla C. Johnson, Janet B.
    Walton, Erin Peters-Burton.
Description: Arlington, VA : National Science Teaching Association, [2020] | Series: STEM road map curriculum
    series | Includes bibliographical references and index.
Identifiers: LCCN 2019029135 (print) | LCCN 2019029136 (ebook) | ISBN 9781681404592 (paperback) |
    ISBN 9781681404608 (pdf)
Subjects: LCSH: Motion--Study and teaching (Early childhood) | Motion--Study and teaching (Preschool) |
    Acceleration (Mechanics)--Study and teaching (Early childhood) | Acceleration (Mechanics)--Study and
    teaching (Preschool) | Rotational motion (Rigid dynamics)--Study and teaching (Early childhood) | Rotational
    motion (Rigid dynamics)--Study and teaching (Preschool) | Physics--Study and teaching (Early childhood) |
    Physics--Study and teaching (Preschool) | Roller coasters. | Kindergarten. | Early childhood education.
Classification: LCC QC128 .P49 2019 (print) | LCC QC128 (ebook) | DDC 372.35--dc23
LC record available at *https://lccn.loc.gov/2019029135*
LC ebook record available at *https://lccn.loc.gov/2019029136*

The *Next Generation Science Standards* ("*NGSS*") were developed by twenty-six states, in collaboration with the National Research Council, the National Science Teaching Association and the American Association for the Advancement of Science in a process managed by Achieve, Inc. For more information go to *www.nextgenscience.org*.

CONTENTS

CONTENTS

ABOUT THE EDITORS AND AUTHORS

Dr. Carla C. Johnson is executive director of the William and Ida Friday Institute for Educational Innovation, associate dean, and professor of science education in the College of Education at North Carolina State University in Raleigh. She was most recently an associate dean, provost fellow, and professor of science education at Purdue University in West Lafayette, Indiana. Dr. Johnson serves as the director of research and evaluation for the Department of Defense–funded Army Educational Outreach Program (AEOP), a global portfolio of STEM education programs, competitions, and apprenticeships. She has been a leader in STEM education for the past decade, serving as the director of STEM Centers, editor of the *School Science and Mathematics* journal, and lead researcher for the evaluation of Tennessee's Race to the Top–funded STEM portfolio. Dr. Johnson has published over 100 articles, books, book chapters, and curriculum books focused on STEM education. She is a former science and social studies teacher and was the recipient of the 2013 Outstanding Science Teacher Educator of the Year award from the Association for Science Teacher Education (ASTE), the 2012 Award for Excellence in Integrating Science and Mathematics from the School Science and Mathematics Association (SSMA), the 2014 award for best paper on Implications of Research for Educational Practice from ASTE, and the 2006 Outstanding Early Career Scholar Award from SSMA. Her research focuses on STEM education policy implementation, effective science teaching, and integrated STEM approaches.

Dr. Janet B. Walton is a senior research scholar and the assistant director of evaluation for AEOP at North Carolina State University's William and Ida Friday Institute for Educational Innovation. She merges her economic development and education backgrounds to develop K–12 curricular materials that integrate real-life issues with sound cross-curricular content. Her research focuses on mixed methods research methodologies and collaboration between schools and community stakeholders for STEM education and problem- and project-based learning pedagogies. With this research agenda, she works to bring contextual STEM experiences into the classroom and provide students and educators with innovative resources and curricular materials.

Dr. Erin Peters-Burton is the Donna R. and David E. Sterling endowed professor in science education at George Mason University in Fairfax, Virginia. She uses her experiences from 15 years as an engineer and secondary science, engineering, and mathematics teacher to develop research projects that directly inform classroom practice in science and engineering. Her research agenda is based on the idea that all students should build self-awareness of how they learn science and engineering. She works to help students see themselves as "science-minded" and help teachers create classrooms that support student skills to develop scientific knowledge. To accomplish this, she pursues research projects that investigate ways that students and teachers can use self-regulated learning theory in science and engineering, as well as how inclusive STEM schools can help students succeed. During her tenure as a secondary teacher, she had a National Board Certification in Early Adolescent Science and was an Albert Einstein Distinguished Educator Fellow for NASA. As a researcher, Dr. Peters-Burton has published over 100 articles, books, book chapters, and curriculum books focused on STEM education and educational psychology. She received the Outstanding Science Teacher Educator of the Year award from ASTE in 2016 and a Teacher of Distinction Award and a Scholarly Achievement Award from George Mason University in 2012, and in 2010 she was named University Science Educator of the Year by the Virginia Association of Science Teachers.

Dr. Andrea R. Milner is the vice president and dean of academic affairs and an associate professor in the Teacher Education Department at Adrian College in Adrian, Michigan. A former early childhood and elementary teacher, Dr. Milner researches the effects constructivist classroom contextual factors have on student motivation and learning strategy use.

Dr. Tamara J. Moore is an associate professor of engineering education in the College of Engineering at Purdue University. Dr. Moore's research focuses on defining STEM integration through the use of engineering as the connection and investigating its power for student learning.

Dr. Vanessa B. Morrison is an associate professor in the Teacher Education Department at Adrian College. She is a former early childhood teacher and reading and language arts specialist whose research is focused on learning and teaching within a transdisciplinary framework.

Dr. Toni A. Sondergeld is an associate professor of assessment, research, and statistics in the School of Education at Drexel University in Philadelphia. Dr. Sondergeld's research concentrates on assessment and evaluation in education, with a focus on K–12 STEM.

ACKNOWLEDGMENTS

This module was developed as a part of the STEM Road Map project (Carla C. Johnson, principal investigator). The Purdue University College of Education, General Motors, and other sources provided funding for this project.

PART 1

THE STEM ROAD MAP
BACKGROUND, THEORY, AND PRACTICE

OVERVIEW OF THE *STEM ROAD MAP CURRICULUM SERIES*

Carla C. Johnson, Erin Peters-Burton, and Tamara J. Moore

The *STEM Road Map Curriculum Series* was conceptualized and developed by a team of STEM educators from across the United States in response to a growing need to infuse real-world learning contexts, delivered through authentic problem-solving pedagogy, into K–12 classrooms. The curriculum series is grounded in integrated STEM, which focuses on the integration of the STEM disciplines—science, technology, engineering, and mathematics—delivered across content areas, incorporating the Framework for 21st Century Learning along with grade-level-appropriate academic standards.

The curriculum series begins in kindergarten, with a five-week instructional sequence that introduces students to the STEM themes and gives them grade-level-appropriate topics and real-world challenges or problems to solve. The series uses project-based and problem-based learning, presenting students with the problem or challenge during the first lesson, and then teaching them science, social studies, English language arts, mathematics, and other content, as they apply what they learn to the challenge or problem at hand.

Authentic assessment and differentiation are embedded throughout the modules. Each *STEM Road Map Curriculum Series* module has a lead discipline, which may be science, social studies, English language arts, or mathematics. All disciplines are integrated into each module, along with ties to engineering. Another key component is the use of STEM Research Notebooks to allow students to track their own learning progress. The modules are designed with a scaffolded approach, with increasingly complex concepts and skills introduced as students progress through grade levels.

The developers of this work view the curriculum as a resource that is intended to be used either as a whole or in part to meet the needs of districts, schools, and teachers who are implementing an integrated STEM approach. A variety of implementation formats are possible, from using one stand-alone module at a given grade level to using all five modules to provide 25 weeks of instruction. Also, within each grade band (K–2, 3–5, 6–8, 9–12), the modules can be sequenced in various ways to suit specific needs.

STANDARDS-BASED APPROACH

The *STEM Road Map Curriculum Series* is anchored in the *Next Generation Science Standards* (*NGSS*), the *Common Core State Standards for Mathematics* (*CCSS Mathematics*), the *Common Core State Standards for English Language Arts* (*CCSS ELA*), and the Framework for 21st Century Learning. Each module includes a detailed curriculum map that incorporates the associated standards from the particular area correlated to lesson plans. The STEM Road Map has very clear and strong connections to these academic standards, and each of the grade-level topics was derived from the mapping of the standards to ensure alignment among topics, challenges or problems, and the required academic standards for students. Therefore, the curriculum series takes a standards-based approach and is designed to provide authentic contexts for application of required knowledge and skills.

THEMES IN THE *STEM ROAD MAP CURRICULUM SERIES*

The K–12 STEM Road Map is organized around five real-world STEM themes that were generated through an examination of the big ideas and challenges for society included in STEM standards and those that are persistent dilemmas for current and future generations:

- Cause and Effect

- Innovation and Progress

- The Represented World

- Sustainable Systems

- Optimizing the Human Experience

These themes are designed as springboards for launching students into an exploration of real-world learning situated within big ideas. Most important, the five STEM Road Map themes serve as a framework for scaffolding STEM learning across the K–12 continuum.

The themes are distributed across the STEM disciplines so that they represent the big ideas in science (Cause and Effect; Sustainable Systems), technology (Innovation and Progress; Optimizing the Human Experience), engineering (Innovation and Progress; Sustainable Systems; Optimizing the Human Experience), and mathematics (The Represented World), as well as concepts and challenges in social studies and 21st century skills that are also excellent contexts for learning in English language arts. The process of developing themes began with the clustering of the *NGSS* performance expectations and the National Academy of Engineering's grand challenges for engineering, which led to the development of the challenge in each module and connections of the module activities to the *CCSS Mathematics* and *CCSS ELA* standards. We performed these mapping processes with large teams of experts and found that these five themes provided breadth, depth, and coherence to frame a high-quality STEM learning experience from kindergarten through 12th grade.

Cause and Effect

The concept of cause and effect is a powerful and pervasive notion in the STEM fields. It is the foundation of understanding how and why things happen as they do. Humans spend considerable effort and resources trying to understand the causes and effects of natural and designed phenomena to gain better control over events and the environment and to be prepared to react appropriately. Equipped with the knowledge of a specific cause-and-effect relationship, we can lead better lives or contribute to the community by altering the cause, leading to a different effect. For example, if a person recognizes that irresponsible energy consumption leads to global climate change, that person can act to remedy his or her contribution to the situation. Although cause and effect is a core idea in the STEM fields, it can actually be difficult to determine. Students should be capable of understanding not only when evidence points to cause and effect but also when evidence points to relationships but not direct causality. The major goal of education is to foster students to be empowered, analytic thinkers, capable of thinking through complex processes to make important decisions. Understanding causality, as well as when it cannot be determined, will help students become better consumers, global citizens, and community members.

Innovation and Progress

One of the most important factors in determining whether humans will have a positive future is innovation. Innovation is the driving force behind progress, which helps create possibilities that did not exist before. Innovation and progress are creative entities, but in the STEM fields, they are anchored by evidence and logic, and they use established concepts to move the STEM fields forward. In creating something new, students must consider what is already known in the STEM fields and apply this knowledge appropriately. When we innovate, we create value that was not there previously and create new conditions and possibilities for even more innovations. Students should consider how their innovations might affect progress and use their STEM thinking to change current human burdens to benefits. For example, if we develop more efficient cars that use by-products from another manufacturing industry, such as food processing, then we have used waste productively and reduced the need for the waste to be hauled away, an indirect benefit of the innovation.

The Represented World

When we communicate about the world we live in, how the world works, and how we can meet the needs of humans, sometimes we can use the actual phenomena to explain a concept. Sometimes, however, the concept is too big, too slow, too small, too fast, or too complex for us to explain using the actual phenomena, and we must use a representation or a model to help communicate the important features. We need representations and models such as graphs, tables, mathematical expressions, and diagrams because it makes our thinking visible. For example, when examining geologic time, we cannot actually observe the passage of such large chunks of time, so we create a timeline or a model that uses a proportional scale to visually illustrate how much time has passed for different eras. Another example may be something too complex for students at a particular grade level, such as explaining the p subshell orbitals of electrons to fifth graders. Instead, we use the Bohr model, which more closely represents the orbiting of planets and is accessible to fifth graders.

When we create models, they are helpful because they point out the most important features of a phenomenon. We also create representations of the world with mathematical functions, which help us change parameters to suit the situation. Creating representations of a phenomenon engages students because they are able to identify the important features of that phenomenon and communicate them directly. But because models are estimates of a phenomenon, they leave out some of the details, so it is important for students to evaluate their usefulness as well as their shortcomings.

Sustainable Systems

From an engineering perspective, the term *system* refers to the use of "concepts of component need, component interaction, systems interaction, and feedback. The interaction of subcomponents to produce a functional system is a common lens used by all engineering disciplines for understanding, analysis, and design." (Koehler, Bloom, and Binns 2013, p. 8). Systems can be either open (e.g., an ecosystem) or closed (e.g., a car battery). Ideally, a system should be sustainable, able to maintain equilibrium without much energy from outside the structure. Looking at a garden, we see flowers blooming, weeds sprouting, insects buzzing, and various forms of life living within its boundaries. This is an example of an ecosystem, a collection of living organisms that survive together, functioning as a system. The interaction of the organisms within the system and the influences of the environment (e.g., water, sunlight) can maintain the system for a period of time, thus demonstrating its ability to endure. Sustainability is a desirable feature of a system because it allows for existence of the entity in the long term.

In the STEM Road Map project, we identified different standards that we consider to be oriented toward systems that students should know and understand in the K–12 setting. These include ecosystems, the rock cycle, Earth processes (such as erosion, tectonics, ocean currents, weather phenomena), Earth-Sun-Moon cycles, heat transfer, and the interaction among the geosphere, biosphere, hydrosphere, and atmosphere. Students and teachers should understand that we live in a world of systems that are not independent of each other, but rather are intrinsically linked such that a disruption in one part of a system will have reverberating effects on other parts of the system.

Optimizing the Human Experience

Science, technology, engineering, and mathematics as disciplines have the capacity to continuously improve the ways humans live, interact, and find meaning in the world, thus working to optimize the human experience. This idea has two components: being more suited to our environment and being more fully human. For example, the progression of STEM ideas can help humans create solutions to complex problems, such as improving ways to access water sources, designing energy sources with minimal impact on our environment, developing new ways of communication and expression, and building efficient shelters. STEM ideas can also provide access to the secrets and wonders of nature. Learning in STEM requires students to think logically and systematically, which is a way of knowing the world that is markedly different from knowing the world as an artist. When students can employ various ways of knowing and understand when it is appropriate to use a different way of knowing or integrate ways of knowing, they are fully experiencing the best of what it is to be human. The problem-based learning scenarios provided in the STEM Road Map help students develop ways of thinking like STEM professionals as they ask questions and design solutions. They learn to optimize the human experience by innovating improvements in the designed world in which they live.

THE NEED FOR AN INTEGRATED STEM APPROACH

At a basic level, STEM stands for science, technology, engineering, and mathematics. Over the past decade, however, STEM has evolved to have a much broader scope and broader implications. Now, educators and policy makers refer to STEM as not only a concentrated area for investing in the future of the United States and other nations but also as a domain and mechanism for educational reform.

The good intentions of the recent decade-plus of focus on accountability and increased testing has resulted in significant decreases not only in instructional time for teaching science and social studies but also in the flexibility of teachers to promote authentic, problem solving–focused classroom environments. The shift has had a detrimental impact on student acquisition of vitally important skills, which many refer to as 21st century skills, and often the ability of students to "think." Further, schooling has become increasingly siloed into compartments of mathematics, science, English language arts, and social studies, lacking any of the connections that are overwhelmingly present in

the real world around children. Students have experienced school as content provided in boxes that must be memorized, devoid of any real-world context, and often have little understanding of why they are learning these things.

STEM-focused projects, curriculum, activities, and schools have emerged as a means to address these challenges. However, most of these efforts have continued to focus on the individual STEM disciplines (predominantly science and engineering) through more STEM classes and after-school programs in a "STEM enhanced" approach (Breiner et al. 2012). But in traditional and STEM enhanced approaches, there is little to no focus on other disciplines that are integral to the context of STEM in the real world. Integrated STEM education, on the other hand, infuses the learning of important STEM content and concepts with a much-needed emphasis on 21st century skills and a problem- and project-based pedagogy that more closely mirrors the real-world setting for society's challenges. It incorporates social studies, English language arts, and the arts as pivotal and necessary (Johnson 2013; Rennie, Venville, and Wallace 2012; Roehrig et al. 2012).

FRAMEWORK FOR STEM INTEGRATION IN THE CLASSROOM

The *STEM Road Map Curriculum Series* is grounded in the Framework for STEM Integration in the Classroom as conceptualized by Moore, Guzey, and Brown (2014) and Moore et al. (2014). The framework has six elements, described in the context of how they are used in the *STEM Road Map Curriculum Series* as follows:

1. The STEM Road Map contexts are meaningful to students and provide motivation to engage with the content. Together, these allow students to have different ways to enter into the challenge.

2. The STEM Road Map modules include engineering design that allows students to design technologies (i.e., products that are part of the designed world) for a compelling purpose.

3. The STEM Road Map modules provide students with the opportunities to learn from failure and redesign based on the lessons learned.

4. The STEM Road Map modules include standards-based disciplinary content as the learning objectives.

5. The STEM Road Map modules include student-centered pedagogies that allow students to grapple with the content, tie their ideas to the context, and learn to think for themselves as they deepen their conceptual knowledge.

6. The STEM Road Map modules emphasize 21st century skills and, in particular, highlight communication and teamwork.

All of the STEM Road Map modules incorporate these six elements; however, the level of emphasis on each of these elements varies based on the challenge or problem in each module.

THE NEED FOR THE *STEM ROAD MAP CURRICULUM SERIES*

As focus is increasing on integrated STEM, and additional schools and programs decide to move their curriculum and instruction in this direction, there is a need for high-quality, research-based curriculum designed with integrated STEM at the core. Several good resources are available to help teachers infuse engineering or more STEM enhanced approaches, but no curriculum exists that spans K–12 with an integrated STEM focus. The next chapter provides detailed information about the specific pedagogy, instructional strategies, and learning theory on which the *STEM Road Map Curriculum Series* is grounded.

REFERENCES

Breiner, J., M. Harkness, C. C. Johnson, and C. Koehler. 2012. What is STEM? A discussion about conceptions of STEM in education and partnerships. *School Science and Mathematics* 112 (1): 3–11.

Johnson, C. C. 2013. Conceptualizing integrated STEM education: Editorial. *School Science and Mathematics* 113 (8): 367–368.

Koehler, C. M., M. A. Bloom, and I. C. Binns. 2013. Lights, camera, action: Developing a methodology to document mainstream films' portrayal of nature of science and scientific inquiry. *Electronic Journal of Science Education* 17 (2).

Moore, T. J., S. S. Guzey, and A. Brown. 2014. Greenhouse design to increase habitable land: An engineering unit. *Science Scope* 37 (7): 51–57.

Moore, T. J., M. S. Stohlmann, H.-H. Wang, K. M. Tank, A. W. Glancy, and G. H. Roehrig. 2014. Implementation and integration of engineering in K–12 STEM education. In *Engineering in pre-college settings: Synthesizing research, policy, and practices*, ed. S. Purzer, J. Strobel, and M. Cardella, 35–60. West Lafayette, IN: Purdue Press.

Rennie, L., G. Venville, and J. Wallace. 2012. *Integrating science, technology, engineering, and mathematics: Issues, reflections, and ways forward*. New York: Routledge.

Roehrig, G. H., T. J. Moore, H. H. Wang, and M. S. Park. 2012. Is adding the *E* enough? Investigating the impact of K–12 engineering standards on the implementation of STEM integration. *School Science and Mathematics* 112 (1): 31–44.

STRATEGIES USED IN THE *STEM ROAD MAP CURRICULUM SERIES*

Erin Peters-Burton, Carla C. Johnson, Toni A. Sondergeld, and Tamara J. Moore

The *STEM Road Map Curriculum Series* uses what has been identified through research as best-practice pedagogy, including embedded formative assessment strategies throughout each module. This chapter briefly describes the key strategies that are employed in the series.

PROJECT- AND PROBLEM-BASED LEARNING

Each module in the *STEM Road Map Curriculum Series* uses either project-based learning or problem-based learning to drive the instruction. Project-based learning begins with a driving question to guide student teams in addressing a contextualized local or community problem or issue. The outcome of project-based instruction is a product that is conceptualized, designed, and tested through a series of scaffolded learning experiences (Blumenfeld et al. 1991; Krajcik and Blumenfeld 2006). Problem-based learning is often grounded in a fictitious scenario, challenge, or problem (Barell 2006; Lambros 2004). On the first day of instruction within the unit, student teams are provided with the context of the problem. Teams work through a series of activities and use open-ended research to develop their potential solution to the problem or challenge, which need not be a tangible product (Johnson 2003).

ENGINEERING DESIGN PROCESS

The *STEM Road Map Curriculum Series* uses engineering design as a way to facilitate integrated STEM within the modules. The engineering design process (EDP) is depicted in Figure 2.1 (p. 10). It highlights two major aspects of engineering design—problem scoping and solution generation—and six specific components of working toward a design: define the problem, learn about the problem, plan a solution, try the solution, test the solution, decide whether the solution is good enough. It also shows that communication

Figure 2.1. Engineering Design Process

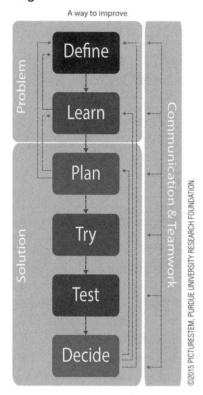

and teamwork are involved throughout the entire process. As the arrows in the figure indicate, the order in which the components of engineering design are addressed depends on what becomes needed as designers progress through the EDP. Designers must communicate and work in teams throughout the process. The EDP is iterative, meaning that components of the process can be repeated as needed until the design is good enough to present to the client as a potential solution to the problem.

Problem scoping is the process of gathering and analyzing information to deeply understand the engineering design problem. It includes defining the problem and learning about the problem. Defining the problem includes identifying the problem, the client, and the end user of the design. The client is the person (or people) who hired the designers to do the work, and the end user is the person (or people) who will use the final design. The designers must also identify the criteria and the constraints of the problem. The criteria are the things the client wants from the solution, and the constraints are the things that limit the possible solutions. The designers must spend significant time learning about the problem, which can include activities such as the following:

- Reading informational texts and researching about relevant concepts or contexts

- Identifying and learning about needed mathematical and scientific skills, knowledge, and tools

- Learning about things done previously to solve similar problems

- Experimenting with possible materials that could be used in the design

Problem scoping also allows designers to consider how to measure the success of the design in addressing specific criteria and staying within the constraints over multiple iterations of solution generation.

Solution generation includes planning a solution, trying the solution, testing the solution, and deciding whether the solution is good enough. Planning the solution includes generating many design ideas that both address the criteria and meet the constraints. Here the designers must consider what was learned about the problem during problem scoping. Design plans include clear communication of design ideas through media such as notebooks, blueprints, schematics, or storyboards. They also include details about the design, such as measurements, materials, colors, costs of materials, instructions for how things fit together, and sets of directions. Making the decision about which design idea to move forward involves considering the trade-offs of each design idea.

Once a clear design plan is in place, the designers must try the solution. Trying the solution includes developing a prototype (a testable model) based on the plan generated. The prototype might be something physical or a process to accomplish a goal. This component of design requires that the designers consider the risk involved in implementing the design. The prototype developed must be tested. Testing the solution includes conducting fair tests that verify whether the plan is a solution that is good enough to meet the client and end user needs and wants. Data need to be collected about the results of the tests of the prototype, and these data should be used to make evidence-based decisions regarding the design choices made in the plan. Here, the designers must again consider the criteria and constraints for the problem.

Using the data gathered from the testing, the designers must decide whether the solution is good enough to meet the client and end user needs and wants by assessment based on the criteria and constraints. Here, the designers must justify or reject design decisions based on the background research gathered while learning about the problem and on the evidence gathered during the testing of the solution. The designers must now decide whether to present the current solution to the client as a possibility or to do more iterations of design on the solution. If they decide that improvements need to be made to the solution, the designers must decide if there is more that needs to be understood about the problem, client, or end user; if another design idea should be tried; or if more planning needs to be conducted on the same design. One way or another, more work needs to be done.

Throughout the process of designing a solution to meet a client's needs and wants, designers work in teams and must communicate to each other, the client, and likely the end user. Teamwork is important in engineering design because multiple perspectives and differing skills and knowledge are valuable when working to solve problems. Communication is key to the success of the designed solution. Designers must communicate their ideas clearly using many different representations, such as text in an engineering notebook, diagrams, flowcharts, technical briefs, or memos to the client.

LEARNING CYCLE

The same format for the learning cycle is used in all grade levels throughout the STEM Road Map, so that students engage in a variety of activities to learn about phenomena in the modules thoroughly and have consistent experiences in the problem- and project-based learning modules. Expectations for learning by younger students are not as high as for older students, but the format of the progression of learning is the same. Students who have learned with curriculum from the STEM Road Map in early grades know

what to expect in later grades. The learning cycle consists of five parts—Introductory Activity/Engagement, Activity/Exploration, Explanation, Elaboration/Application of Knowledge, and Evaluation/Assessment—and is based on the empirically tested 5E model from BSCS (Bybee et al. 2006).

In the Introductory Activity/Engagement phase, teachers introduce the module challenge and use a unique approach designed to pique students' curiosity. This phase gets students to start thinking about what they already know about the topic and begin wondering about key ideas. The Introductory Activity/Engagement phase positions students to be confident about what they are about to learn, because they have prior knowledge, and clues them into what they don't yet know.

In the Activity/Exploration phase, the teacher sets up activities in which students experience a deeper look at the topics that were introduced earlier. Students engage in the activities and generate new questions or consider possibilities using preliminary investigations. Students work independently, in small groups, and in whole-group settings to conduct investigations, resulting in common experiences about the topic and skills involved in the real-world activities. Teachers can assess students' development of concepts and skills based on the common experiences during this phase.

During the Explanation phase, teachers direct students' attention to concepts they need to understand and skills they need to possess to accomplish the challenge. Students participate in activities to demonstrate their knowledge and skills to this point, and teachers can pinpoint gaps in student knowledge during this phase.

In the Elaboration/Application of Knowledge phase, teachers present students with activities that engage in higher-order thinking to create depth and breadth of student knowledge, while connecting ideas across topics within and across STEM. Students apply what they have learned thus far in the module to a new context or elaborate on what they have learned about the topic to a deeper level of detail.

In the last phase, Evaluation/Assessment, teachers give students summative feedback on their knowledge and skills as demonstrated through the challenge. This is not the only point of assessment (as discussed in the section on Embedded Formative Assessments), but it is an assessment of the culmination of the knowledge and skills for the module. Students demonstrate their cognitive growth at this point and reflect on how far they have come since the beginning of the module. The challenges are designed to be multidimensional in the ways students must collaborate and communicate their new knowledge.

STEM RESEARCH NOTEBOOK

One of the main components of the *STEM Road Map Curriculum Series* is the STEM Research Notebook, a place for students to capture their ideas, questions, observations, reflections, evidence of progress, and other items associated with their daily work. At the beginning of each module, the teacher walks students through the setup of the STEM

Research Notebook, which could be a three-ring binder, composition book, or spiral notebook. You may wish to have students create divided sections so that they can easily access work from various disciplines during the module. Electronic notebooks kept on student devices are also acceptable and encouraged. Students will develop their own table of contents and create chapters in the notebook for each module.

Each lesson in the *STEM Road Map Curriculum Series* includes one or more prompts that are designed for inclusion in the STEM Research Notebook and appear as questions or statements that the teacher assigns to students. These prompts require students to apply what they have learned across the lesson to solve the big problem or challenge for that module. Each lesson is designed to meaningfully refer students to the larger problem or challenge they have been assigned to solve with their teams. The STEM Research Notebook is designed to be a key formative assessment tool, as students' daily entries provide evidence of what they are learning. The notebook can be used as a mechanism for dialogue between the teacher and students, as well as for peer and self-evaluation.

The use of the STEM Research Notebook is designed to scaffold student notebooking skills across the grade bands in the *STEM Road Map Curriculum Series*. In the early grades, children learn how to organize their daily work in the notebook as a way to collect their products for future reference. In elementary school, students structure their notebooks to integrate background research along with their daily work and lesson prompts. In the upper grades (middle and high school), students expand their use of research and data gathering through team discussions to more closely mirror the work of STEM experts in the real world.

THE ROLE OF ASSESSMENT IN THE *STEM ROAD MAP CURRICULUM SERIES*

Starting in the middle years and continuing into secondary education, the word *assessment* typically brings grades to mind. These grades may take the form of a letter or a percentage, but they typically are used as a representation of a student's content mastery. If well thought out and implemented, however, classroom assessment can offer teachers, parents, and students valuable information about student learning and misconceptions that does not necessarily come in the form of a grade (Popham 2013).

The *STEM Road Map Curriculum Series* provides a set of assessments for each module. Teachers are encouraged to use assessment information for more than just assigning grades to students. Instead, assessments of activities requiring students to actively engage in their learning, such as student journaling in STEM Research Notebooks, collaborative presentations, and constructing graphic organizers, should be used to move student learning forward. Whereas other curriculum with assessments may include objective-type (multiple-choice or matching) tests, quizzes, or worksheets, we have intentionally avoided these forms of assessments to better align assessment strategies with teacher instruction and

student learning techniques. Since the focus of this book is on project- or problem-based STEM curriculum and instruction that focuses on higher-level thinking skills, appropriate and authentic performance assessments were developed to elicit the most reliable and valid indication of growth in student abilities (Brookhart and Nitko 2008).

Comprehensive Assessment System

Assessment throughout all STEM Road Map curriculum modules acts as a comprehensive system in which formative and summative assessments work together to provide teachers with high-quality information on student learning. Formative assessment occurs when the teacher finds out formally or informally what a student knows about a smaller, defined concept or skill and provides timely feedback to the student about his or her level of proficiency. Summative assessments occur when students have performed all activities in the module and are given a cumulative performance evaluation in which they demonstrate their growth in learning.

A comprehensive assessment system can be thought of as akin to a sporting event. Formative assessments are the practices: It is important to accomplish them consistently, they provide feedback to help students improve their learning, and making mistakes can be worthwhile if students are given an opportunity to learn from them. Summative assessments are the competitions: Students need to be prepared to perform at the best of their ability. Without multiple opportunities to practice skills along the way through formative assessments, students will not have the best chance of demonstrating growth in abilities through summative assessments (Black and Wiliam 1998).

Embedded Formative Assessments

Formative assessments in this module serve two main purposes: to provide feedback to students about their learning and to provide important information for the teacher to inform immediate instructional needs. Providing feedback to students is particularly important when conducting problem- or project-based learning because students take on much of the responsibility for learning, and teachers must facilitate student learning in an informed way. For example, if students are required to conduct research for the Activity/Exploration phase but are not familiar with what constitutes a reliable resource, they may develop misconceptions based on poor information. When a teacher monitors this learning through formative assessments and provides specific feedback related to the instructional goals, students are less likely to develop incomplete or incorrect conceptions in their independent investigations. By using formative assessment to detect problems in student learning and then acting on this information, teachers help move student learning forward through these teachable moments.

Formative assessments come in a variety of formats. They can be informal, such as asking students probing questions related to student knowledge or tasks or simply observing students engaged in an activity to gather information about student skills. Formative assessments can also be formal, such as a written quiz or a laboratory practical. Regardless of the type, three key steps must be completed when using formative assessments (Sondergeld, Bell, and Leusner 2010). First, the assessment is delivered to students so that teachers can collect data. Next, teachers analyze the data (student responses) to determine student strengths and areas that need additional support. Finally, teachers use the results from information collected to modify lessons and create learning environments that reinforce weak points in student learning. If student learning information is not used to modify instruction, the assessment cannot be considered formative in nature.

Formative assessments can be about content, science process skills, or even learning skills. When a formative assessment focuses on content, it assesses student knowledge about the disciplinary core ideas from the *Next Generation Science Standards* (*NGSS*) or content objectives from *Common Core State Standards for Mathematics* (*CCSS Mathematics*) or *Common Core State Standards for English Language Arts* (*CCSS ELA*). Content-focused formative assessments ask students questions about declarative knowledge regarding the concepts they have been learning. Process skills formative assessments examine the extent to which a student can perform science and engineering practices from the *NGSS* or process objectives from *CCSS Mathematics* or *CCSS ELA*, such as constructing an argument. Learning skills can also be assessed formatively by asking students to reflect on the ways they learn best during a module and identify ways they could have learned more.

Assessment Maps

Assessment maps or blueprints can be used to ensure alignment between classroom instruction and assessment. If what students are learning in the classroom is not the same as the content on which they are assessed, the resultant judgment made on student learning will be invalid (Brookhart and Nitko 2008). Therefore, the issue of instruction and assessment alignment is critical. The assessment map for this book (found in Chapter 3) indicates by lesson whether the assessment should be completed as a group or on an individual basis, identifies the assessment as formative or summative in nature, and aligns the assessment with its corresponding learning objectives.

Note that the module includes far more formative assessments than summative assessments. This is done intentionally to provide students with multiple opportunities to practice their learning of new skills before completing a summative assessment. Note also that formative assessments are used to collect information on only one or two learning objectives at a time so that potential relearning or instructional modifications can focus on smaller and more manageable chunks of information. Conversely, summative assessments in the module cover many more learning objectives, as they are traditionally used as final markers of student learning. This is not to say that information collected from summative assessments cannot or should not be used formatively. If teachers find that gaps in student

learning persist after a summative assessment is completed, it is important to revisit these existing misconceptions or areas of weakness before moving on (Black et al. 2003).

SELF-REGULATED LEARNING THEORY IN THE STEM ROAD MAP MODULES

Many learning theories are compatible with the STEM Road Map modules, such as constructivism, situated cognition, and meaningful learning. However, we feel that the self-regulated learning theory (SRL) aligns most appropriately (Zimmerman 2000). SRL requires students to understand that thinking needs to be motivated and managed (Ritchhart, Church, and Morrison 2011). The STEM Road Map modules are student centered and are designed to provide students with choices, concrete hands-on experiences, and opportunities to see and make connections, especially across subjects (Eliason and Jenkins 2012; NAEYC 2016). Additionally, SRL is compatible with the modules because it fosters a learning environment that supports students' motivation, enables students to become aware of their own learning strategies, and requires reflection on learning while experiencing the module (Peters and Kitsantas 2010).

The theory behind SRL (see Figure 2.2) explains the different processes that students engage in before, during, and after a learning task. Because SRL is a cyclical learning process, the accomplishment of one cycle develops strategies for the next learning cycle. This cyclic way of learning aligns with the various sections in the STEM Road Map lesson plans on Introductory Activity/Engagement, Activity/Exploration, Explanation, Elaboration/Application of Knowledge, and Evaluation/Assessment. Since the students engaged in a module take on much of the responsibility for learning, this theory also provides guidance for teachers to keep students on the right track.

The remainder of this section explains how SRL theory is embedded within the five sections of each module and points out ways to

Figure 2.2. SRL Theory

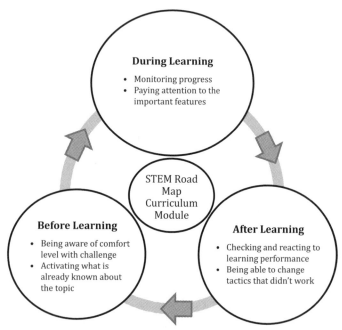

Source: Adapted from Zimmerman 2000.

support students in becoming independent learners of STEM while productively functioning in collaborative teams.

Before Learning: Setting the Stage

Before attempting a learning task such as the STEM Road Map modules, teachers should develop an understanding of their students' level of comfort with the process of accomplishing the learning and determine what they already know about the topic. When students are comfortable with attempting a learning task, they tend to take more risks in learning and as a result achieve deeper learning (Bandura 1986).

The STEM Road Map curriculum modules are designed to foster excitement from the very beginning. Each module has an Introductory Activity/Engagement section that introduces the overall topic from a unique and exciting perspective, engaging the students to learn more so that they can accomplish the challenge. The Introductory Activity also has a design component that helps teachers assess what students already know about the topic of the module. In addition to the deliberate designs in the lesson plans to support SRL, teachers can support a high level of student comfort with the learning challenge by finding out if students have ever accomplished the same kind of task and, if so, asking them to share what worked well for them.

During Learning: Staying the Course

Some students fear inquiry learning because they aren't sure what to do to be successful (Peters 2010). However, the STEM Road Map curriculum modules are embedded with tools to help students pay attention to knowledge and skills that are important for the learning task and to check student understanding along the way. One of the most important processes for learning is the ability for learners to monitor their own progress while performing a learning task (Peters 2012). The modules allow students to monitor their progress with tools such as the STEM Research Notebooks, in which they record what they know and can check whether they have acquired a complete set of knowledge and skills. The STEM Road Map modules support inquiry strategies that include previewing, questioning, predicting, clarifying, observing, discussing, and journaling (Morrison and Milner 2014). Through the use of technology throughout the modules, inquiry is supported by providing students access to resources and data while enabling them to process information, report the findings, collaborate, and develop 21st century skills.

It is important for teachers to encourage students to have an open mind about alternative solutions and procedures (Milner and Sondergeld 2015) when working through the STEM Road Map curriculum modules. Novice learners can have difficulty knowing what to pay attention to and tend to treat each possible avenue for information as equal (Benner 1984). Teachers are the mentors in a classroom and can point out ways for students to approach learning during the Activity/Exploration, Explanation, and

Elaboration/Application of Knowledge portions of the lesson plans to ensure that students pay attention to the important concepts and skills throughout the module. For example, if a student is to demonstrate conceptual awareness of motion when working on roller coaster research, but the student has misconceptions about motion, the teacher can step in and redirect student learning.

After Learning: Knowing What Works

The classroom is a busy place, and it may often seem that there is no time for self-reflection on learning. Although skipping this reflective process may save time in the short term, it reduces the ability to take into account things that worked well and things that didn't so that teaching the module may be improved next time. In the long run, SRL skills are critical for students to become independent learners who can adapt to new situations. By investing the time it takes to teach students SRL skills, teachers can save time later, because students will be able to apply methods and approaches for learning that they have found effective to new situations. In the Evaluation/Assessment portion of the STEM Road Map curriculum modules, as well as in the formative assessments throughout the modules, two processes in the after-learning phase are supported: evaluating one's own performance and accounting for ways to adapt tactics that didn't work well. Students have many opportunities to self-assess in formative assessments, both in groups and individually, using the rubrics provided in the modules.

The designs of the *NGSS* and *CCSS* allow for students to learn in diverse ways, and the STEM Road Map curriculum modules emphasize that students can use a variety of tactics to complete the learning process. For example, students can use STEM Research Notebooks to record what they have learned during the various research activities. Notebook entries might include putting objectives in students' own words, compiling their prior learning on the topic, documenting new learning, providing proof of what they learned, and reflecting on what they felt successful doing and what they felt they still needed to work on. Perhaps students didn't realize that they were supposed to connect what they already knew with what they learned. They could record this and would be prepared in the next learning task to begin connecting prior learning with new learning.

SAFETY IN STEM

Student safety is a primary consideration in all subjects but is an area of particular concern in science, where students may interact with unfamiliar tools and materials that may pose additional safety risks. It is important to implement safety practices within the context of STEM investigations, whether in a classroom laboratory or in the field. When you keep safety in mind as a teacher, you avoid many potential issues with the lesson while also protecting your students.

STEM safety practices encompass things considered in the typical science classroom. Ensure that students are familiar with basic safety considerations, such as wearing protective equipment (e.g., safety glasses or goggles and latex-free gloves) and taking care with sharp objects, and know emergency exit procedures. Teachers should learn beforehand the locations of the safety eyewash, fume hood, fire extinguishers, and emergency shut-off switch in the classroom and how to use them. Also be aware of any school or district safety policies that are in place and apply those that align with the work being conducted in the lesson. It is important to review all safety procedures annually.

STEM investigations should always be supervised. Each lesson in the modules includes teacher guidelines for applicable safety procedures that should be followed. Before each investigation, teachers should go over these safety procedures with the student teams. Some STEM focus areas such as engineering require that students can demonstrate how to properly use equipment in the maker space before the teacher allows them to proceed with the lesson.

The National Science Teaching Association (NSTA) provides a list of science rules and regulations, including standard operating procedures for lab safety, and a safety acknowledgment form for students and parents or guardians to sign. You can access these resources at *http://static.nsta.org/pdfs/SafetyInTheScienceClassroom.pdf.* In addition, NSTA's Safety in the Science Classroom web page (*www.nsta.org/safety*) has numerous links to safety resources, including papers written by the NSTA Safety Advisory Board.

Disclaimer: The safety precautions for each activity are based on use of the recommended materials and instructions, legal safety standards, and better professional practices. Using alternative materials or procedures for these activities may jeopardize the level of safety and therefore is at the user's own risk.

REFERENCES

Bandura, A. 1986. *Social foundations of thought and action: A social cognitive theory.* Englewood Cliffs, NJ: Prentice-Hall.

Barell, J. 2006. *Problem-based learning: An inquiry approach.* Thousand Oaks, CA: Corwin Press.

Benner, P. 1984. *From novice to expert: Excellence and power in clinical nursing practice.* Menlo Park, CA: Addison-Wesley.

Black, P., C. Harrison, C. Lee, B. Marshall, and D. Wiliam. 2003. *Assessment for learning: Putting it into practice.* Berkshire, UK: Open University Press.

Black, P., and D. Wiliam. 1998. Inside the black box: Raising standards through classroom assessment. *Phi Delta Kappan* 80 (2): 139–148.

Blumenfeld, P., E. Soloway, R. Marx, J. Krajcik, M. Guzdial, and A. Palincsar. 1991. Motivating project-based learning: Sustaining the doing, supporting learning. *Educational Psychologist* 26 (3): 369–398.

Brookhart, S. M., and A. J. Nitko. 2008. *Assessment and grading in classrooms.* Upper Saddle River, NJ: Pearson.

Bybee, R., J. Taylor, A. Gardner, P. Van Scotter, J. Carlson Powell, A. Westbrook, and N. Landes. 2006. *The BSCS 5E instructional model: Origins and effectiveness*. Colorado Springs, CO: BSCS.

Eliason, C. F., and L. T. Jenkins. 2012. *A practical guide to early childhood curriculum*. 9th ed. New York: Merrill.

Johnson, C. 2003. Bioterrorism is real-world science: Inquiry-based simulation mirrors real life. *Science Scope* 27 (3): 19–23.

Krajcik, J., and P. Blumenfeld. 2006. Project-based learning. In *The Cambridge handbook of the learning sciences*, ed. R. Keith Sawyer, 317–334. New York: Cambridge University Press.

Lambros, A. 2004. *Problem-based learning in middle and high school classrooms: A teacher's guide to implementation*. Thousand Oaks, CA: Corwin Press.

Milner, A. R., and T. Sondergeld. 2015. Gifted urban middle school students: The inquiry continuum and the nature of science. *National Journal of Urban Education and Practice* 8 (3): 442–461.

Morrison, V., and A. R. Milner. 2014. Literacy in support of science: A closer look at cross-curricular instructional practice. *Michigan Reading Journal* 46 (2): 42–56.

National Association for the Education of Young Children (NAEYC). 2016. Developmentally appropriate practice position statements. *www.naeyc.org/positionstatements/dap*.

Peters, E. E. 2010. Shifting to a student-centered science classroom: An exploration of teacher and student changes in perceptions and practices. *Journal of Science Teacher Education* 21 (3): 329–349.

Peters, E. E. 2012. Developing content knowledge in students through explicit teaching of the nature of science: Influences of goal setting and self-monitoring. *Science and Education* 21 (6): 881–898.

Peters, E. E., and A. Kitsantas. 2010. The effect of nature of science metacognitive prompts on science students' content and nature of science knowledge, metacognition, and self-regulatory efficacy. *School Science and Mathematics* 110: 382–396.

Popham, W. J. 2013. *Classroom assessment: What teachers need to know*. 7th ed. Upper Saddle River, NJ: Pearson.

Ritchhart, R., M. Church, and K. Morrison. 2011. *Making thinking visible: How to promote engagement, understanding, and independence for all learners*. San Francisco, CA: Jossey-Bass.

Sondergeld, T. A., C. A. Bell, and D. M. Leusner. 2010. Understanding how teachers engage in formative assessment. *Teaching and Learning* 24 (2): 72–86.

Zimmerman, B. J. 2000. Attaining self-regulation: A social-cognitive perspective. In *Handbook of self-regulation*, ed. M. Boekaerts, P. Pintrich, and M. Zeidner, 13–39. San Diego: Academic Press.

PART 2

PHYSICS IN MOTION

STEM ROAD MAP MODULE

PHYSICS IN MOTION MODULE OVERVIEW

Vanessa B. Morrison, Andrea R. Milner, Janet B. Walton, Carla C. Johnson, and Erin Peters-Burton

THEME: Cause and Effect

LEAD DISCIPLINE: Science

MODULE SUMMARY

This module uses roller coasters as an entry point for students to explore the physics of motion. Students work collaboratively to investigate concepts such as energy, gravity, friction, and speed. Students use the engineering design process (EDP) as they create and evaluate their own mini roller coasters in the classroom (adapted from Koehler, Bloom, and Milner, 2015).

ESTABLISHED GOALS AND OBJECTIVES

At the conclusion of this module, students will be able to do the following:

- Demonstrate awareness of concepts associated with motion, energy, gravity, force, push, pull, speed, inertia, direction, slope, and friction through play

- Use technology to gather research information and communicate

- Measure, compare, and evaluate numbers related to module concepts

- Demonstrate awareness of concepts associated with motion by discussing, investigating, and creating marble track roller coasters

- Identify careers associated with roller coaster design and construction

- Describe and apply the EDP

- Design, construct, test, and evaluate marble track roller coasters

CHALLENGE OR PROBLEM FOR STUDENTS TO SOLVE: ROLLER COASTER DESIGN CHALLENGE

Student teams are challenged to create a marble track roller coaster that meets specific design criteria. Students investigate, design, construct, test, and evaluate their tracks and decide on the best design.

CONTENT STANDARDS ADDRESSED IN THIS STEM ROAD MAP MODULE

A full listing with descriptions of the standards this module addresses can be found in Appendix C (p. 111). Listings of the particular standards addressed within lessons are provided in a table for each lesson in Chapter 4.

STEM RESEARCH NOTEBOOK

Each student should maintain a STEM Research Notebook, which will serve as a place for students to organize their work throughout this module (see p. 12 for more general discussion on setup and use of the notebook). All written work in the module should be included in the notebook, including records of students' thoughts and ideas, fictional accounts based on the concepts in the module, and records of student progress through the EDP. The notebooks may be maintained across subject areas, giving students the opportunity to see that although their classes may be separated during the school day, the knowledge they gain is connected. The lesson plans for this module contain STEM Research Notebook Entry sections (numbered 1–14), and templates for each notebook entry are included in Appendix A (p. 93).

Emphasize to students the importance of organizing all information in a Research Notebook. Explain to them that scientists and other researchers maintain detailed Research Notebooks in their work. These notebooks, which are crucial to researchers' work because they contain critical information and track the researchers' progress, are often considered legal documents for scientists who are pursuing patents or wish to provide proof of their discovery process.

MODULE LAUNCH

Launch the module by conducting an interactive read-aloud of *Energy in Motion,* by Melissa Stewart. Next, have students participate in a small movement investigation and then view the video "Sid the Science Kid: 'Sid's Super Kick,' part 2," found at *www.dailymotion.com/video/x15oaoe.*

PREREQUISITE SKILLS FOR THE MODULE

Students enter this module with a wide range of preexisting skills, information, and knowledge. Table 3.1 provides an overview of prerequisite skills and knowledge that students are expected to apply in this module, along with examples of how they apply this knowledge throughout the module. Differentiation strategies are also provided for students who may need additional support in acquiring or applying this knowledge.

Table 3.1. Prerequisite Key Knowledge and Examples of Applications and Differentiation Strategies

Prerequisite Key Knowledge	Application of Knowledge by Students	Differentiation for Students Needing Additional Support
Science • Understand cause and effect	*Science* • Determine how specific design elements of a marble track influence the marble's behavior on the track.	*Science* • Provide students with content via books, videos, songs, and computer programs to help students understand the motion of roller coasters and other objects affected by gravity. • Read aloud picture books to class, and have students identify cause and effect sequences.
Mathematics • Number sense	*Mathematics* • Use comparative measurements to make decisions to enhance the construction of marble tracks.	*Mathematics* • Provide examples of ways to measure observed phenomena such as distance, height, and time. • Model measurement techniques using standard and nonstandard units of measurement. • Read aloud nonfiction texts about measurement to class. • Provide opportunities for students to practice measurement in a variety of settings (e.g., in the classroom and outdoors).

Continued

Table 3.1. (*continued*)

Prerequisite Key Knowledge	Application of Knowledge by Students	Differentiation for Students Needing Additional Support
Language and Inquiry Skills • Visualize • Make predictions • Record ideas and observations using pictures and words • Ask and respond to questions	*Language and Inquiry Skills* • Make and confirm or reject predictions. • Share thought processes through notebooking, asking and responding to questions, and using the engineering design process.	*Language and Inquiry Skills* • As a class, make predictions when reading fictional texts. • As a class, make predictions about observed natural phenomena (e.g., gravity). • Model the process of using information and prior knowledge to make predictions. • Provide samples of notebook entries.
Speaking and Listening • Participate in group discussions	*Speaking and Listening* • Engage in collaborative group discussions in the development of the marble tracks.	*Speaking and Listening* • Model speaking and listening skills. • Create a class list of good listening and good speaking practices. • Read aloud picture books that feature collaboration and teamwork.

POTENTIAL STEM MISCONCEPTIONS

Students enter the classroom with a wide variety of prior knowledge and ideas, so it is important to be alert to misconceptions, or inappropriate understandings of foundational knowledge. These misconceptions can be classified as one of several types: "preconceived notions," opinions based on popular beliefs or understandings; "nonscientific beliefs," knowledge students have gained about science from sources outside the scientific community; "conceptual misunderstandings," incorrect conceptual models based on incomplete understanding of concepts; "vernacular misconceptions," misunderstandings of words based on their common use versus their scientific use; and "factual misconceptions," incorrect or imprecise knowledge learned in early life that remains unchallenged (NRC 1997, p. 28). Misconceptions must be addressed and dismantled for students to reconstruct their knowledge, and therefore teachers should be prepared to take the following steps:

- *Identify students' misconceptions.*

- *Provide a forum for students to confront their misconceptions.*

- *Help students reconstruct and internalize their knowledge, based on scientific models. (NRC 1997, p. 29)*

Keeley and Harrington (2010) recommend using diagnostic tools such as probes and formative assessment to identify and confront student misconceptions and begin the process of reconstructing student knowledge. Keeley's *Uncovering Student Ideas in Science* series contains probes targeted toward uncovering student misconceptions in a variety of areas and may be a useful resource for addressing student misconceptions in this module.

Some commonly held misconceptions specific to lesson content are provided with each lesson so that you can be alert for student misunderstanding of the science concepts presented and used during this module. The American Association for the Advancement of Science has also identified misconceptions that students frequently hold regarding various science concepts (see the links at *http://assessment.aaas.org/topics*).

SRL PROCESS COMPONENTS

Table 3.2 (p. 28) illustrates some of the activities in the Physics in Motion module and how they align with the self-regulated learning (SRL) process before, during, and after learning.

Table 3.2. SRL Process Components

Learning Process Components	Example From Physics in Motion Module	Lesson Number and Learning Component
BEFORE LEARNING		
Motivates students	Students perform fun activities such as hopping on one leg and waving their hands to investigate what they know about motion. Students also participate in an interactive read-aloud of the book *Energy in Motion*.	Lesson 1, Introductory Activity/Engagement
Evokes prior learning	Students demonstrate their own knowledge about motion and things that move through play in their STEM Research Notebooks.	Lesson 1, Introductory Activity/Engagement
DURING LEARNING		
Focuses on important features	Students use the engineering design process (EDP) to design virtual roller coasters. Students are guided through the process using the steps of the EDP by addressing specific questions in the lesson. Teachers track student responses to the EDP questions.	Lesson 2, Activity/ Exploration
Helps students monitor their progress	Students share their most successful virtual roller coaster designs with the class, by showing either their designs on the computer or their sketches of their best designs and explaining the designs.	Lesson 2, Activity/ Exploration
AFTER LEARNING		
Evaluates learning	Student teams share the marble track roller coasters they designed, built, tested, and improved in the lesson. The class discusses the physics concepts relevant to the success or failure of track designs.	Lesson 3, Explanation
Takes account of what worked and what did not work	Students design additional marble tracks incorporating what they learned from their own and other teams' designs.	Lesson 3, Elaboration/ Application of Knowledge

STRATEGIES FOR DIFFERENTIATING INSTRUCTION WITHIN THIS MODULE

For the purposes of this curriculum module, differentiated instruction is conceptualized as a way to tailor instruction—including process, content, and product—to various student needs in your class. A number of differentiation strategies are integrated into lessons across the module. The problem- and project-based learning approach used in the lessons is designed to address students' multiple intelligences by providing a variety of entry points and methods to investigate the key concepts in the module (e.g., investigating motion through play, through structured investigations, and through interactive read-alouds). Differentiation strategies for students needing support in prerequisite knowledge can be found in Table 3.1 (p. 25). You are encouraged to use information gained about student prior knowledge during introductory activities and discussions to inform your instructional differentiation. Strategies incorporated into this lesson include flexible grouping, varied environmental learning contexts, assessments, compacting, tiered assignments and scaffolding, and mentoring.

Flexible Grouping. Students work collaboratively in a variety of activities throughout this module. Grouping strategies you might employ include student-led grouping, grouping students according to ability level or common interests, grouping students randomly, or grouping them so that students in each group have complementary strengths (for instance, one student might be strong in mathematics, another in art, and another in writing).

Varied Environmental Learning Contexts. Students have the opportunity to learn in various contexts throughout the module, including alone, in groups, in quiet reading and research-oriented activities, and in active learning in inquiry and design activities. In addition, students learn in a variety of ways, including through doing inquiry activities, journaling, reading a variety of texts, watching videos, participating in class discussion, and conducting web-based research.

Assessments. Students are assessed in a variety of ways throughout the module, including individual and collaborative formative and summative assessments. Students have the opportunity to produce work via written text, oral and media presentations, and modeling.

Compacting. Based on student prior knowledge you may wish to adjust instructional activities for students who exhibit prior mastery of a learning objective. Since student work in science is largely collaborative throughout the module, this strategy may be most appropriate for mathematics, social studies, or ELA activities. You may wish to compile a classroom database of supplementary readings for a variety of reading levels and on a variety of topics related to the module's topic to provide opportunities for students to undertake independent reading.

Tiered Assignments and Scaffolding. Based on your awareness of student ability, understanding of concepts, and mastery of skills, you may wish to provide students with variations on activities by adding complexity to assignments or providing more or fewer learning supports for activities throughout the module. For instance, some students may need additional support in reading or may benefit from cloze sentence handouts to enhance vocabulary understanding. Other students may benefit from expanded reading selections and additional reflective writing or from working with manipulatives and other visual representations of mathematical concepts. You may also work with your school librarian to compile a set of topical resources at a variety of reading levels.

Mentoring. As group design teamwork becomes increasingly complex throughout the module, you may wish to have a resource teacher, older student, or volunteer work with groups that struggle to stay on-task and collaborate effectively.

STRATEGIES FOR ENGLISH LANGUAGE LEARNERS

Students who are developing proficiency in English language skills require additional supports to simultaneously learn academic content and the specialized language associated with specific content areas. WIDA (2012) has created a framework for providing support to these students and makes available rubrics and guidance on differentiating instructional materials for English language learners (ELLs). In particular, ELL students may benefit from additional sensory supports such as images, physical modeling, and graphic representations of module content, as well as interactive support through collaborative work. This module incorporates a variety of sensory supports and offers ongoing opportunities for ELL students to work collaboratively.

When differentiating instruction for ELL students, you should carefully consider the needs of these students as you introduce and use academic language in various language domains (listening, speaking, reading, and writing) throughout this module. To adequately differentiate instruction for ELL students, you should have an understanding of the proficiency level of each student. The following five overarching preK–5 WIDA learning standards are relevant to this module:

- Standard 1: Social and Instructional Language. Focus on social behavior in group work and class discussions, following directions, and information gathering.

- Standard 2: The Language of Language Arts. Tell a story or recount an experience with appropriate facts and relevant, descriptive details, speaking audibly in coherent sentences.

- Standard 3: The Language of Mathematics. Order three objects by length; compare the lengths of two objects indirectly by using a third object. Analyze text of word problems.

- Standard 4: The Language of Science. Focus on safety practices, energy sources, ecology and conservation, natural resources, and scientific inquiry.

- Standard 5: The Language of Social Studies. Focus on resources and products; needs of groups, societies, and cultures; and location of objects and places.

SAFETY CONSIDERATIONS FOR THE ACTIVITIES IN THIS MODULE

The safety precautions associated with each investigation are based in part on the use of the recommended materials and instructions, legal safety standards, and better professional safety practices. Selection of alternative materials or procedures for these investigations may jeopardize the level of safety and therefore is at the user's own risk. Remember that an investigation includes three parts: (1) setup, in which you prepare the materials for students to use; (2) the actual hands-on investigation, in which students use the materials and equipment; and (3) cleanup, in which you or the students clean the materials and put them away for later use. The safety procedures for each investigation apply to all three parts. For more general safety guidelines, see the Safety in STEM section in Chapter 2 (p. 18).

We also recommend that you go over the safety rules that are included as part of the safety acknowledgment form with your students before beginning the first investigation. Once you have gone over these rules with your students, have them sign the safety acknowledgment form. You should also send the form home with students for parents or guardians to read and sign to acknowledge that they understand the safety procedures that must be followed by their children. A sample elementary safety acknowledgment form can be found on the NSTA Safety Portal at *http://static.nsta.org/pdfs/ SafetyAcknowledgmentForm-ElementarySchool.pdf*.

DESIRED OUTCOMES AND MONITORING SUCCESS

The desired outcome for this module is outlined in Table 3.3, along with suggested ways to gather evidence to monitor student success. For more specific details on desired outcomes, see the Established Goals and Objectives sections for the module and individual lessons.

Table 3.3. Desired Outcome and Evidence of Success in Achieving Identified Outcome

Desired Outcome	Evidence of Success	
	Performance Tasks	Other Measures
Students understand and can demonstrate concepts associated with the physics of motion. Students apply these concepts in their marble track designs.	• Student teams design, construct, test, and evaluate multiple styles of roller coaster tracks. • Students each maintain a STEM Research Notebook with responses to questions and observations.	Students are assessed using the Observation, STEM Research Notebook, and Participation Rubric.

ASSESSMENT PLAN OVERVIEW AND MAP

Table 3.4 provides an overview of the major group and individual *products* and *deliverables,* or things that student teams will produce in this module, that constitute the assessment for this module. See Table 3.5 for a full assessment map of formative and summative assessments in this module.

Table 3.4. Major Products and Deliverables in Lead Disciplines for Groups and Individuals

Lesson	Major Group Products and Deliverables	Major Individual Products and Deliverables
1	• Team participation in Physics in Motion Game Days • Team participation in Playground Pals	• STEM Research Notebook entries #1–8
2	• Team virtual roller coaster designs and presentations	• STEM Research Notebook entries #9–12
3	• Team marble track roller coaster designs and presentations	• STEM Research Notebook entries #13–14

Table 3.5. Assessment Map for Physics in Motion Module

Lesson	Assessment	Group/ Individual	Formative/ Summative	Lesson Objective Assessed
1	STEM Research Notebook *entries*	Individual	Formative	• Predict how toys will move when students apply various forces. • Predict how the equipment and their bodies will move when they use playground equipment. • Observe how toys move when students apply various forces. • Observe how the equipment and their bodies move when they use playground equipment. • Compare and contrast the motion of toys and the motion of playground equipment and their bodies. • Identify the forces of push and pull.
1	Physics in Motion Game Days *activity*	Group	Formative	• Predict how toys will move when students apply various forces. • Observe how toys move when students apply various forces. • Identify the forces of push and pull. • Identify gravity as a force.
1	Playground Pals *activity*	Group	Formative	• Predict how the equipment and their bodies will move when they use playground equipment. • Observe how the equipment and their bodies move when they use playground equipment. • Compare and contrast the motion of toys and the motion of playground equipment and their bodies. • Identify the forces of push and pull. • Identify gravity as a force.
2	STEM Research Notebook *entries*	Individual	Formative	• Describe how a roller coaster works. • Identify what safety precautions might be important for roller coaster riders. • Use the engineering design process (EDP) to create and evaluate virtual roller coaster tracks. • Communicate and present findings about virtual roller coaster tracks.

Continued

Table 3.5. (*continued*)

Lesson	Assessment	Group/ Individual	Formative/ Summative	Lesson Objective Assessed
2	Design Time! *activity*	Group	Formative	• Understand how a roller coaster works. • Identify and explain gravity as a force that works on roller coaster cars. • Communicate and present findings about virtual roller coaster tracks.
3	Roller Coaster Design Challenge *activity*	Group	Summative	• Use the EDP to create marble track roller coasters. • Communicate and present findings about their marble track roller coasters. • Use their understanding of safety to create safety guidelines for their roller coasters. • Use their understanding of roller coasters to create flyers about their roller coasters.

MODULE TIMELINE

Tables 3.6–3.10 (pp. 35–37) provide lesson timelines for each week of the module. These timelines are provided for general guidance only and are based on class times of approximately 30 minutes.

Table 3.6. STEM Road Map Module Schedule for Week One

Day 1	Day 2	Day 3	Day 4	Day 5
Lesson 1 *Let's Explore Motion Through Play!* • Launch the module and introduce challenge. • Introduce STEM Research Notebooks. • Introduce motion with a class discussion.	*Lesson 1* *Let's Explore Motion Through Play!* • Hold a class discussion about motion and things that move. • Conduct interactive read-aloud of *Give It a Push! Give It a Pull! A Look at Forces*, by Jennifer Boothroyd.	*Lesson 1* *Let's Explore Motion Through Play!* • Introduce motion of kicking as a way of pushing by showing "Sid the Science Kid: 'Sid's Super Kick,' part 2" video. • Discuss motion of kicking and record student learning.	*Lesson 1* *Let's Explore Motion Through Play!* • Conduct interactive read-aloud of *Energy in Motion*, by Melissa Stewart. • Start class vocabulary chart.	*Lesson 1* *Let's Explore Motion Through Play!* • Introduce Physics in Motion Game Days. • Introduce engineering as a career. • Create class list of good team habits *(optional)*.

Table 3.7. STEM Road Map Module Schedule for Week Two

Day 6	Day 7	Day 8	Day 9	Day 10
Lesson 1 *Let's Explore Motion Through Play!* • Prepare for Physics in Motion Game Days by making and recording predictions about motion.	*Lesson 1* *Let's Explore Motion Through Play!* • Have students make and record observations of Physics in Motion Game Days activity.	*Lesson 1* *Let's Explore Motion Through Play!* • Continue Physics in Motion Game Days activity by having students experiment with different heights of ramps.	*Lesson 1* *Let's Explore Motion Through Play!* • Conduct interactive read-aloud of *And Everyone Shouted, "Pull!" A First Look at Forces and Motion*, by Claire Llewellyn. • Continue Physics in Motion Game Days activity by comparing students' predictions and observations.	*Lesson 1* *Let's Explore Motion Through Play!* • Conclude Physics in Motion Game Days activity by having students offer explanations for their observations. • Conduct interactive read-aloud of *Gravity*, by Jason Chin. • Introduce measurement.

Table 3.8. STEM Road Map Module Schedule for Week Three

Day 11	Day 12	Day 13	Day 14	Day 15
Lesson 1 *Let's Explore Motion Through Play!* • Conduct interactive read-aloud of the poem "The Seesaw" from *Poems to Count On*, by Sandra Liatsos. • Introduce Playground Pals activity and have students make predictions.	*Lesson 1* *Let's Explore Motion Through Play!* • Conclude Playground Pals activity by having students make observations.	*Lesson 1* *Let's Explore Motion Through Play!* • Have students offer explanations for their Playground Pals activity observations. • Conduct a lesson assessment.	*Lesson 1* *Let's Explore Motion Through Play!* • Discuss safety guidelines for playgrounds. • Conduct interactive read-aloud of *I Can Be Safe: A First Look at Safety*, by Pat Thomas.	*Lesson 2* *Roller Coaster Fun!* • Hold a class discussion about roller coasters. • Show video of roller coaster. • Conduct interactive read-aloud of *Roller Coaster*, by Marla Frazee.

Table 3.9. STEM Road Map Module Schedule for Week Four

Day 16	Day 17	Day 18	Day 19	Day 20
Lesson 2 *Roller Coaster Fun!* • Introduce size and speed comparisons to describe roller coasters. • Introduce careers associated with roller coasters. • Conduct interactive read-aloud of *Archibald Frisby*, by Michael Chesworth.	*Lesson 2* *Roller Coaster Fun!* • Show video about students designing a roller coaster. • Introduce the engineering design process. • Introduce the Design Time! virtual roller coaster design activity.	*Lesson 2* *Roller Coaster Fun!* • Have student teams design their roller coasters for the Design Time! activity. • Have student teams present their designs.	*Lesson 2* *Roller Coaster Fun!* • Discuss students' observations of their virtual roller coasters. • Conduct interactive read-aloud of *Gravity in Action: Roller Coasters!* by Joan Newton.	*Lesson 2* *Roller Coaster Fun!* • Discuss roller coaster safety guidelines. • Hold a class discussion on environmental impacts of roller coasters. • Introduce concept of speed.

Table 3.10. STEM Road Map Module Schedule for Week Five

Day 21	Day 22	Day 23	Day 24	Day 25
Lesson 3 Roller Coaster Design Challenge	*Lesson 3 Roller Coaster Design Challenge*	*Lesson 3 Roller Coaster Design Challenge*	*Lesson 3 Roller Coaster Design Challenge*	*Lesson 3 Roller Coaster Design Challenge*
• Hold a class discussion about the engineering design process. • Review the module challenge and materials. • Introduce requirements for teams' roller coasters.	• Review science content in *Gravity in Action: Roller Coasters!* by Joan Newton. • Begin Roller Coaster Design Challenge (Define and Learn).	• Continue Roller Coaster Design Challenge (Plan, Try, and Test). • Students create safety rules for their roller coasters.	• Complete Roller Coaster Design Challenge (Decide). • Teams share roller coaster designs.	• Students build additional marble tracks incorporating what they learned from their own and other teams' designs. • Teams create a plan for a roller coaster to be placed within a theme park. • Each student creates a flyer for the team's roller coaster.

RESOURCES

The media specialist can help teachers locate resources for students to view and read about roller coasters and related physics content. Special educators and reading specialists can help find supplemental sources for students needing extra support in reading and writing. Additional resources may be found online. Community resources for this module may include civil engineers and mechanical engineers.

REFERENCES

Keeley, P., and R. Harrington. 2010. *Uncovering student ideas in physical science, volume 1: 45 new force and motion assessment probes.* Arlington, VA: NSTA Press.

Koehler, C., M. A. Bloom, and A. R. Milner. 2015. The STEM Road Map for grades K–2. In *STEM Road Map: A framework for integrated STEM education,* ed. C. C. Johnson, E. E. Peters-Burton, and T. J. Moore, 41–67. New York: Routledge. *www.routledge.com/products/9781138804234.*

National Research Council (NRC). 1997. *Science teaching reconsidered: A handbook.* Washington, DC: National Academies Press.

WIDA. 2012. 2012 amplification of the English language development standards: Kindergarten–grade 12. *https://wida.wisc.edu/teach/standards/eld.*

PHYSICS IN MOTION LESSON PLANS

Vanessa B. Morrison, Andrea R. Milner, Janet B. Walton, Carla C. Johnson, and Erin Peters-Burton

Lesson Plan 1: Let's Explore Motion Through Play!

In this lesson, students are introduced to the module challenge and begin to explore the concept of physics in motion through everyday play experiences. Students make predictions about the motion of toys and participate in Physics in Motion Game Days, during which they test those predictions. Students also make predictions about motion related to using playground equipment and test these predictions during a visit to a playground.

ESSENTIAL QUESTIONS

- How do things move?

- What causes things to move?

ESTABLISHED GOALS AND OBJECTIVES

At the conclusion of this lesson, students will be able to do the following:

- Predict how toys will move when students apply various forces

- Predict how the equipment and their bodies will move when they use playground equipment

- Observe how toys move when students apply various forces

- Observe how the equipment and their bodies move when they use playground equipment

- Compare and contrast the motion of toys and the motion of playground equipment and their bodies

- Identify the forces of push and pull

- Identify gravity as a force

TIME REQUIRED

- 14 days (approximately 30 minutes each day; see Tables 3.6–3.8, pp. 35–36)

MATERIALS

Required Materials for Lesson 1

- STEM Research Notebooks
- Computer with internet access for videos
- Masking tape
- Tape measure or yardstick
- Stopwatch
- Scale
- Books
  - *And Everyone Shouted, "Pull!" A First Look at Forces and Motion*, by Claire Llewellyn (Hodder Wayland, 2001)
  - *Energy in Motion*, by Melissa Stewart (Children's Press, 2006)
  - *Give It a Push! Give It a Pull! A Look at Forces*, by Jennifer Boothroyd (Lerner, 2010)
  - *Gravity*, by Jason Chin (Roaring Brook Press, 2014)
  - *I Can Be Safe: A First Look at Safety*, by Pat Thomas (B.E.S. Publishing, 2003)
  - *Poems to Count On*, by Sandra Liatsos (Scholastic Teaching Resources, 1999)
- Chart paper
- Markers
- Safety glasses with side shields or safety goggles

Additional Materials for Physics in Motion Game Days (per team of 3–4)

- 2 wheeled rolling toys less than 6 inches wide, 1 larger than the other
- 1 small ball (e.g., golf ball, table tennis ball)
- 1 small marble
- 1 piece of ¼ inch thick balsa wood (about 6 × 24 inches)

- Materials to elevate balsa wood ramp about 12 and 18 inches from the floor (e.g., plastic shoe boxes or stacks of books)

Additional Materials for Playground Pals (per pair of students)

- Piece of carpet (12–15 inches square)

- Piece of corrugated cardboard (12–15 inches square)

- Piece of waxed paper (12–15 inches square)

SAFETY NOTES

1. All students must wear safety glasses with side shields or goggles during all phases of this inquiry activity.

2. Direct supervision is required during all aspects of this activity to make sure safety behaviors are followed and enforced.

3. Make sure any items dropped on the floor or ground are picked up to avoid slip- or trip-and-fall hazards.

4. Tell students to be careful when handling wood to avoid splinters.

5. Make sure all fragile materials, furniture, and equipment are out of the activity area before students experiment with applying forces to objects.

6. Have students wash hands with soap and water after completing this activity.

CONTENT STANDARDS AND KEY VOCABULARY

Table 4.1 (p. 42) lists the content standards from the *Next Generation Science Standards* (*NGSS*), *Common Core State Standards* (*CCSS*), National Association for the Education of Young Children (NAEYC), and the Framework for 21st Century Learning that this lesson addresses, and Table 4.2 (p. 45) presents the key vocabulary. Vocabulary terms are provided for both teacher and student use. Teachers may choose to introduce some or all of the terms to students.

Table 4.1. Content Standards Addressed in STEM Road Map Module
Lesson 1

NEXT GENERATION SCIENCE STANDARDS

PERFORMANCE EXPECTATIONS

- K-PS2-1. Plan and conduct an investigation to compare the effects of different strengths or different directions of pushes and pulls on the motion of an object.

- K-PS2-2. Analyze data to determine if a design solution works as intended to change the speed or direction of an object with a push or a pull.

SCIENCE AND ENGINEERING PRACTICE

Planning and Carrying Out Investigations

- With guidance, plan and conduct an investigation in collaboration with peers.

DISCIPLINARY CORE IDEAS

PS2.A. Forces and Motion

- Pushes and pulls can have different strengths and directions.

- Pushing or pulling on an object can change the speed or direction of its motion and can start or stop it.

PS2.B. Types of Interactions

- When objects touch or collide, they push on one another and can change motion.

PS3.C. Relationship Between Energy and Forces

- A bigger push or pull makes things speed up or slow down more quickly.

CROSSCUTTING CONCEPT

Cause and Effect

- Simple tests can be designed to gather evidence to support or refute student ideas about causes.

COMMON CORE STATE STANDARDS FOR MATHEMATICS

MATHEMATICAL PRACTICES

- MP1. Make sense of problems and persevere in solving them.

- MP2. Reason abstractly and quantitatively.

- MP3. Construct viable arguments and critique the reasoning of others.

- MP4. Model with mathematics.

- MP5. Use appropriate tools strategically.

- MP6. Attend to precision.

Continued

Table 4.1. (*continued*)

MATHEMATICAL CONTENT

- K.MD.B.3. Classify objects into given categories; count the numbers of objects in each category and sort the categories by count.

- K.CC.C.6. Identify whether the number of objects in one group is greater than, less than, or equal to the number of objects in another group, e.g., by using matching and counting strategies.

- K.CC.C.7. Compare two numbers between 1 and 10 presented as written numerals.

- K.CC.B.4. Understand the relationship between numbers and quantities; connect counting to cardinality.

- K.CC.B.4a. When counting objects, say the number names in the standard order, pairing each object with one and only one number name and each number name with one and only one object.

- K.CC.B.4b. Understand that the last number name said tells the number of objects counted. The number of objects is the same regardless of their arrangement or the order in which they were counted.

- K.CC.B.4c. Understand that each successive number name refers to a quantity that is one larger.

- K.MD.A.1. Describe measurable attributes of objects, such as length or weight. Describe several measurable attributes of a single object.

- K.MD.A.2. Directly compare two objects with a measurable attribute in common, to see which object has "more of"/"less of" the attribute, and describe the difference. For example, directly compare the heights of two children and describe one child as taller/shorter.

COMMON CORE STATE STANDARDS FOR ENGLISH LANGUAGE ARTS

READING STANDARDS

- RI.K.1. With prompting and support, ask and answer questions about key details in a text.

- RI.K.3. With prompting and support, describe the connection between two individuals, events, ideas, or pieces of information in a text.

WRITING STANDARDS

- W.K.2. Use a combination of drawing, dictating, and writing to compose informative/explanatory texts in which they name what they are writing about and supply some information about the topic.

- W.K.5. With guidance and support from adults, respond to questions and suggestions from peers and add details to strengthen writing as needed.

- W.K.7. Participate in shared research and writing projects (e.g., explore a number of books by a favorite author and express opinions about them).

Continued

Table 4.1. (*continued*)

SPEAKING AND LISTENING STANDARDS

- SL.K.1. Participate in collaborative conversations with diverse partners about *kindergarten topics and texts* with peers and adults in small and larger groups.

- SL.K.3. Ask and answer questions in order to seek help, get information, or clarify something that is not understood.

- SL.K.5. Add drawings or other visual displays to descriptions as desired to provide additional detail.

NATIONAL ASSOCIATION FOR THE EDUCATION OF YOUNG CHILDREN STANDARDS

- 2.G.02. Children are provided varied opportunities and materials to learn key content and principles of science.

- 2.G.03. Children are provided with varied opportunities and materials that encourage them to use the five senses to observe, explore, and experiment with scientific phenomena.

- 2.G.04. Children are provided with varied opportunities to use simple tools to observe objects and scientific phenomena.

- 2.G.05. Children are provided with varied opportunities and materials to collect data and to represent and document their findings (e.g., through drawing or graphing).

- 2.G.06. Children are provided with varied opportunities and materials that encourage them to think, question, and reason about observed and inferred phenomena.

- 2.G.07. Children are provided with varied opportunities and materials that encourage them to discuss scientific concepts in everyday conversation.

- 2.G.08. Children are provided with varied opportunities and materials that help them learn and use scientific terminology and vocabulary associated with the content areas.

- 2.H.03. Technology is used to extend learning within the classroom and integrate and enrich the curriculum.

FRAMEWORK FOR 21ST CENTURY LEARNING

- Interdisciplinary Themes; Learning and Innovation Skills; Information, Media, and Technology Skills; Life and Career Skills

---

Table 4.2. Key Vocabulary for Lesson 1

Key Vocabulary	Definition
direction	the path something takes
energy	what causes objects to change and move
force	a push or pull on an object
gravity	the force of attraction that causes things to fall toward Earth
measurement	a way to describe the size, shape, or motion of something using numbers
motion	the act of changing location or position
pull	to move something toward or with you
push	to move something away from you
speed	how fast something moves

TEACHER BACKGROUND INFORMATION

Kindergartners are rapidly developing across all domains (physical, social and emotional, personality, cognitive, and language). They are beginning to develop autonomy, logical thinking skills, and the ability to reason out problems. Throughout this module, you should support and facilitate the advancement of these content areas and developmental domains within each student. For information about how formative assessments can be used to connect student prior experiences with classroom instruction, see the STEM Teaching Tools resource "Making Science Instruction Compelling for All Students: Using Cultural Formative Assessment to Build on Learner Interest and Experience" at *http://stemteachingtools.org/pd/sessionc.*

Motion and Forces

Motion, or the act of changing location or position, is a natural part of our daily lives. The study of motion, or mechanics, is a central focus of the field of physics. Motion is a result of various forces acting on matter. Forces cause objects to begin to move, accelerate, and slow or stop moving once in motion. Isaac Newton (1643–1727), a British scientist, mathematician, and astronomer, studied forces and motion and is credited with discovering the force of gravity. Newton also developed laws of motion that are the foundation for much of our modern understanding of mechanics. For more information about Isaac Newton, see the following websites:

- *www.biography.com/news/how-isaac-newton-changed-our-world*

- *www.ducksters.com/biography/scientists/isaac_newton.php*

Forces are vector quantities, meaning that they have both a magnitude (measured in units of newtons, or the force necessary to move a 1 kilogram object at an acceleration of 1 meter per second squared) and direction. Newton's first law of motion states that an object at rest will remain at rest until another force acts on it and that an object in motion will remain in motion until another force acts on it. Force is expressed mathematically as the product of an object's mass, or how much matter the object contains, and acceleration, or an object's change in speed ($F = m \times a$). This is Newton's second law, and it means that an object with a greater mass requires stronger forces to move at the same acceleration as an object with a smaller mass.

There are a variety of forces at work in our world that can be broadly categorized as contact forces or as distance forces. Contact forces are those that require making physical contact with an object to affect its motion. These include applied force, frictional force, normal force, tension force, and spring force. Distance forces are those that work over a distance and do not require making physical contact with an object. These include gravitational force, electrical force, and magnetic force. The focus of this module is on applied, frictional, and gravitational forces, since these are among the most readily observable. Applied force, or force applied by a person or object, is presented as pushes or pulls within the module. Frictional force, or the force applied by a surface as an object moves against it, is dependent on the material of the surface and of the object (e.g., we slide more easily on ice than on pavement). Gravitational force is the force exerted by a very large object (e.g., Earth or the Moon), which results in other objects being attracted toward it. For the purposes of this module, gravitational force will be conceptualized as the downward pull of Earth on objects. For more information about forces, see the following websites:

- *www.physics4kids.com/files/motion_intro.html*

- *www.physicsclassroom.com/class/newtlaws/Lesson-2/Types-of-Forces*

While much of the science and mathematics of motion are beyond the scope of kindergarten, students experience the physics of motion every day. From kicking soccer balls to pumping their legs on a swing, they see and experience motion at work. Students can observe that applying forces allows us to change our own or objects' positions and obtain specific motions. This module provides students with opportunities to learn about the physics of motion through investigating toys, playground equipment, and roller coasters.

Career Connections

You may wish to introduce careers associated with engineers and other STEM workers during this module, such as the following (adapted from Koehler, Bloom, and Milner 2015):

- engineer

- geographer

- journalist

- mathematician

- physicist

- environmental scientist

For more information about these and other careers, see the Bureau of Labor Statistics' *Occupational Outlook Handbook* at *www.bls.gov/ooh/home.html*.

In this module, students are introduced to the idea that engineers and other STEM workers work together in teams to solve problems. Students experience working in teams and in pairs as they progress through a simple scientific process including predicting, observing, and explaining phenomena related to forces in this lesson. This introduction to teamwork sets the stage for students' use of the engineering design process (EDP) later in the module.

Know, Learning, Evidence, Wonder, Scientific Principles (KLEWS) Charts

You will track student knowledge on Know, Learning, Evidence, Wonder, Scientific Principles (KLEWS) charts throughout this module. These charts will be used to access and assess student prior knowledge, encourage students to think critically about the topic under discussion, and track student learning throughout the module. Using KLEWS charts challenges students to connect evidence and scientific principles with their learning. Be sure to list the topic at the top of each chart. The charts should consist of five columns—one for each KLEWS component. It may be helpful to post these charts in a prominent place in the classroom so that students can refer to them throughout the module. Students will include their personal ideas and reflections in their STEM Research Notebook entries. For more information about KLEWS charts, see the January 2006 National Science Teaching Association *WebNews Digest* article "Evidence Helps the KWL get a KLEW" at *www.nsta.org/publications/news/story.aspx?id=51519* or the February 2015 *Science and Children* article "Methods and Strategies: KLEWS to Explanation-Building in Science" at *www.nsta.org/store/product_detail.aspx?id=10.2505/4/sc15_052_06_66*.

Interactive Read-Alouds

This module also uses interactive read-alouds to engage students, access their prior knowledge, develop student background knowledge, and introduce topical vocabulary. These read-alouds expose children to teacher-read literature that may be beyond their independent reading levels but is consistent with their listening level. Interactive read-alouds may incorporate a variety of techniques, and you can find helpful information regarding these techniques at the following websites:

- *www.readingrockets.org/article/repeated-interactive-read-alouds-preschool-and-kindergarten*

- *www.k5chalkbox.com/interactive-read-aloud.html*

- *www.readwritethink.org/professional-development/strategy-guides/teacher-read-aloud-that-30799.html*

In general, interactive read-alouds provide opportunities for students to share prior knowledge and experiences, interact with the text and concepts introduced therein, launch conversations about the topics introduced, construct meaning, make predictions, and draw comparisons. You may wish to mark places within the texts to pause and ask for students to share experiences, predictions, or other ideas. Each reading experience should focus on an ongoing interaction between students and the text, making time to do the following:

- Allow students to share personal stories throughout the reading.

- Ask students to predict throughout the story.

- Allow students to add new ideas from the book to the KLEWS chart and their STEM Research Notebooks.

- Allow students to add new words from the book to the vocabulary chart and their STEM Research Notebooks.

The materials list for each lesson includes the books for interactive read-alouds that you will use in that lesson. A list of suggested books for additional reading can be found at the end of this chapter (see p. 86).

Predict, Observe, Explain (POE)

Students use a predict, observe, explain (POE) sequence in their investigations in this module. This strategy develops young learners' inquiry skills in a way that reflects the core components of the scientific process of asking questions, formulating hypotheses, conducting experiments, making observations, and drawing conclusions or offering new hypotheses. In the POE sequence, students are presented with a scenario and challenged to make a prediction (e.g., what will happen when a ball is released at the top of a hill?). Next, students are given an opportunity to observe the scenario (e.g., releasing balls onto various inclines). Students record their observations, offer explanations, and revisit their predictions to compare them with the actual outcome. For more information about the POE sequence, see "Predict, Observe, Explain (POE)" at *https://arbs.nzcer.org.nz/predict-observe-explain-poe*.

COMMON MISCONCEPTIONS

Students will have various types of prior knowledge about the concepts introduced in this lesson. Table 4.3 outlines some common misconceptions students may have concerning these concepts. Because of the breadth of students' experiences, it is not possible to anticipate every misconception that students may bring as they approach this lesson. Incorrect or inaccurate prior understanding of concepts can influence student learning in the future, however, so it is important to be alert to misconceptions such as those presented in the table.

Table 4.3. Common Misconceptions About the Concepts in Lesson 1

Topic	Student Misconception	Explanation
Forces: A force is a push or pull between objects	Nonmoving objects (for example, a tabletop or roadway) do not exert a force.	Stationary objects can exert forces on other objects. For example, when you roll a toy car over a tabletop, there is a frictional force between the table and the car wheels.
	If a moving object is slowing, the force that was propelling it forward is decreasing.	When a force acts on a moving object in a direction opposite the direction of motion, the moving object will slow, even if the force that was propelling the object forward continues.

PREPARATION FOR LESSON 1

Review the Teacher Background Information section (p. 45), assemble the materials for the lesson, duplicate the student handouts, and preview the video recommended in the Learning Components section that follows. Present students with their STEM Research Notebooks and explain how these will be used (see p. 24). Templates for the STEM Research Notebook are provided in Appendix A (p. 93), and a rubric for observations, student participation, and STEM Research Notebook entries is provided in Appendix B (p. 109).

STEM Research Notebook Entry #4 provides a template for students to record vocabulary words. You may wish to use this template throughout the module for students to record definitions and illustrations of key vocabulary words. The template provides space for definitions and illustrations of three words. If you introduce more than three vocabulary words in a lesson, you should make multiple copies of the template for each student.

In this lesson, students have the opportunity to predict, observe, and explain physics in motion via play. Ask students to bring toys from home that exhibit motion without the use of batteries or electronics for the Physics in Motion Game Days activity. You should

also be prepared to provide additional items for this activity so that each team of three or four students has access to the listed materials (p. 40). Prepare for the activity by marking a starting and finish line for each team by placing a 12-inch-long piece of masking tape on the floor to serve as the starting line and another piece 3 feet away to serve as the finish line.

Plan for students to visit the school playground or another playground for the Playground Pals activity (p. 56). For students to complete the activity as described, the playground should have swings, slides, and a seesaw. If it does not, you should be prepared to adapt the activity to the available playground equipment.

LEARNING COMPONENTS
Introductory Activity/Engagement

Connection to the Challenge: Begin each day of this lesson by directing students' attention to the module challenge, the Roller Coaster Design Challenge. Introduce the challenge by telling students the following:

> *Our class is going to learn about how people and things move. One thing that moves in exciting ways is a roller coaster ride at an amusement park. After we learn about how things move, you will work with a team to design and build an exciting roller coaster for a make-believe rider (a marble).*

To begin the module, hold a class discussion about what teams will need to know to design and build roller coasters. Hold a brief class discussion each day thereafter of how students' learning in the previous days' lessons contributed to their ability to complete the challenge. You may wish to create a class list of key ideas on chart paper.

Science Class and Mathematics and ELA Connections: Introduce students to the STEM Research Notebooks that they will use throughout this module and provide guidelines for their use. Emphasize to students that professional scientists and engineers use research notebooks to keep track of their thoughts and ideas and to record their observations; students will use their notebooks in the same way.

Introduce the concept of motion. Have students demonstrate movement. For example, ask students to stand up and hop on their left legs and then on their right legs, then to wave their left hands and then their right hands.

Hold a classroom discussion about motion. Following agreed-upon rules for discussions, ask your students the following:

- What are some things that can move?

- How can you describe how they move? (e.g., fast [car], slow [baby crawling], up [throwing a ball], down [going down a slide], bouncing [ball])

Document student responses on a KLEWS chart, then have students record their ideas about motion in STEM Research Notebook Entry #1.

STEM Research Notebook Entry #1

Have students document their own ideas about things that they can move and how these things move in their STEM Research Notebooks, using both words and pictures.

Review students' responses. Ask students to identify things that move fast and things that move slow. Hold a class discussion about how students know that some things move quickly and other things move more slowly (e.g., by watching, by seeing how far things travel in the same amount of time). Use an example of a baby crawling and a car driving on a highway. Ask students to describe how fast these things move, introducing the concept of speed. Ask them to give examples of animals that move slowly and animals that move faster, recording students' answers on a chart with two columns, one labeled "slow" and one labeled "fast." After students have named several animals, ask them to name other things that move slowly and fast.

Next, ask students to share their ideas about how they can make things move (e.g., by throwing, by kicking). Record students' ideas on the KLEWS chart. Demonstrate pushing a chair forward. Ask students to name what sort of motion you used (pushing). Ask them to name examples of some things they can move by pushing. Document student responses on the KLEWS chart.

Then, demonstrate pulling by pulling the chair back toward you. Ask students to name what sort of motion you used (pulling). Ask them to name some examples of things they can pull. Document student responses on the KLEWS chart.

Conduct an interactive read-aloud of *Give It a Push! Give It a Pull! A Look at Forces*, by Jennifer Boothroyd. Document student ideas about what they learned about forces and motion on the KLEWS chart, then have students complete STEM Research Notebook Entry #2.

STEM Research Notebook Entry #2

Have students document what they learned about forces and motion in their STEM Research Notebooks, using both words and pictures.

Social Studies Connection: Help students connect their examples of motion with careers. Examples might include a car moving (delivery driver, race-car driver, taxi driver); throwing (baseball player, sanitation worker); lifting (daycare worker, warehouse worker). List students' answers on chart paper (keep up throughout the module).

Activity/Exploration

Science Class and Mathematics and ELA Connections: Introduce the motion of kicking as a way of pushing by showing the video "Sid the Science Kid: 'Sid's Super Kick,' part 2," found at *www.dailymotion.com/video/x15oaoe*. Before showing the video, ask students to share their ideas about the motion of kicking and their experiences with kicking, documenting student responses on a KLEWS chart. After showing the video, ask them to share what they learned about kicking, adding to the KLEWS chart and having students complete STEM Research Notebook Entry #3.

STEM Research Notebook Entry #3

After students watch the video, have them document what they learned about kicking in their STEM Research Notebooks, using both words and pictures.

Next, conduct an interactive read-aloud of *Energy in Motion*, by Melissa Stewart, asking students to listen for "energy words" throughout the reading. With the words students identify, start a class vocabulary chart with words and pictures. Post the chart on the classroom wall for reference in class discussions, and add vocabulary to the chart throughout the remainder of the module.

STEM Research Notebook Entry #4

Have students add vocabulary words and definitions in their STEM Research Notebooks, using both words and pictures.

Physics in Motion Game Days

Introduce the Physics in Motion Game Days activity by telling students that they are going to work in teams to explore how toys move. Introduce the idea that engineers are people who solve problems using science, mathematics, and technology and that engineers often work in teams to solve problems. In this lesson, the problem their team is going to solve is identifying what makes toys move. You may wish to have students create a class list of good team habits (e.g., making sure everyone on the team has a chance to do all the activities, using kind words with other team members, not talking when another team member is talking).

Tell students that part of what scientists do is to make predictions. Develop a class definition of the word *prediction,* and have students share some ways that they predict things (e.g., predicting the weather, predicting what gift they might receive for their birthday, predicting how a person might respond when they speak to them).

Next, tell students that they will make predictions about the ways that toys will move in response to the team members' actions. Show students a rolling toy and demonstrate different ways to push the toy forward (e.g., flicking the toy with your middle finger, blowing on the toy, pushing it gently with your hand, and pushing it vigorously with your hand).

STEM Research Notebook Entry #5

Ask students the following questions, having them record their predictions about the behavior of their toys on the floor and on the ramp in their STEM Research Notebooks:

When I push the toy with a medium push:

- In what direction will the toy move? (forward, backward, to the right, or to the left)
- How far will the toy move? (it will not reach the finish line, it will reach the finish line, or it will move past the finish line)

When I flick the toy with one finger:

- How far will the toy move? (it will not reach the finish line, it will reach the finish line, or it will move past the finish line)

When I blow on the toy:

- How far will the toy move? (it will not reach the finish line, it will reach the finish line, or it will move past the finish line)

Then, ask:

- Will there be differences in movement between the bigger and smaller toys when I push them with a medium push?

Show students a small ball, a toy car, a marble, and one of the balsa wood ramps elevated 12 inches above the floor and ask:

- Will the ball, car, and marble need a push to move down the ramp?
- Will the toys move faster on the higher or lower ramp?

After students document their predictions in response to each question in their STEM Research Notebooks, have students share their predictions and document class predictions on chart paper, using a POE chart such as the one in Figure 4.1 (p. 54).

Figure 4.1. Sample POE Chart

Predict	Observe	Explain

STEM Research Notebook Entry #6

Have students make observations about the behavior of their toys on the floor and on the ramp and record their observations in their STEM Research Notebooks.

Ask students to think about what scientists and engineers might do as they test their predictions. Guide students to understand that scientists and engineers observe things and record what they observe. As a class, have students work to generate ideas for how they should make and record their observations. Working together as a class, fill in the Observe column of the POE chart and their STEM Research Notebook entries. Have student teams each choose one wheeled toy to use to make observations for the first four questions. For question 4, have students use both the larger and smaller wheeled toys for their observations.

After students have completed observations for the first four questions, allow them to experiment with various types of forces on the toys. For example, you may wish to direct students to do the following:

- Push the toy away from them from the marked starting line to the finish line. Ask students what amount of force—gentle, medium, or strong—is necessary to move the toy from the starting line to the finish line, but not beyond. Is it different for different toys?

- Pull the toy toward them. Ask students how this is different from pushing it.

Next, demonstrate to students how to set up their ramps with one end on the plastic boxes or books so that it is about 12 inches from the floor and how to adjust the ramps to a taller height (about 18 inches) by adding books or boxes to the stack. Demonstrate to students how to release a toy down the ramp without pushing it. Then, have students make observations for questions 5 and 6 on STEM Research Notebook Entry #6.

After students have recorded their observations, have them experiment with various toys on ramps of different heights. Instruct students to roll a small ball, a toy car, and a marble down the lower ramp, asking students which toy seems to go fastest and which seems to go slowest. Next, have students elevate their ramps to the higher height and repeat their experiment with the toys. You may wish to pair teams to hold "races" by having one team elevate its ramp to the lower height and the other team elevate its ramp to the higher height, and then release each kind of toy simultaneously to see which ramp the toys move faster on.

Social Studies Connection: Not applicable.

Explanation

Science Class and Mathematics and ELA Connections: Have students review their predictions from before the Physics in Motion Game Days. Ask students whether their predictions about how their toys moved in response to their pushes were right or not and why. Have them share their ideas about why they think their toys responded as they did to their pushes (questions 1–4 on STEM Research Notebook entries #5 and #6). Use students' ideas to fill in the Explain column of the POE chart.

Next, conduct an interactive read-aloud of *And Everyone Shouted, "Pull!" A First Look at Forces and Motion,* by Claire Llewellyn. Focus students' attention on how wheels help things move. After the read-aloud, ask students to share any additional ideas they have about why their toys responded the way they did to their pushes in the Physics in Motion Game Days activity.

Then, ask students whether their predictions about how their toys moved on the ramps were right or not and why. Have them share their ideas about why they think their toys responded as they did to their pushes (questions 5–7 on STEM Research Notebook entries #5 and #6). Add students' ideas to the Explain column of the POE chart.

Next, introduce the concept of gravity by conducting an interactive read-aloud of *Gravity,* by Jason Chin. After the read-aloud, ask students to share any additional ideas they have about why their toys behaved the way they did on the ramps in the Physics in Motion Game Days activity.

Remind students about how they made observations about the motion of their toys (they compared the motion caused by various forces on different toys). Ask if they can think of other ways to describe the distance their toys moved. Introduce the concept of measurement as a way of assigning numbers to describe how things move. Ask students for their ideas about how they could assign numbers to their observations, recording students' ideas on a class chart. Introduce the idea that scientists and engineers use special tools to take measurements. Show students a measuring tape or yardstick and ask how they could use this to measure the motion of their toys (measure the distance they moved). Then, show students a stopwatch and ask how they could use this to measure the motion of their toys (measure the amount of time it takes for a toy to move to the finish line or down the ramp). Finally, show students a scale and ask what they could measure with this (the weight of each of the toys).

Social Studies Connection: Not applicable.

Elaboration/Application of Knowledge

Science Class and ELA and Mathematics Connections: Students explore forces and motion on the playground in the Playground Pals activity. Conduct an interactive read-aloud of the poem "The Seesaw" from *Poems to Count On,* by Sandra Liatsos. Following agreed-upon rules for discussions, ask your students how weight and size affected the hippo's ability to play on a seesaw. This will help students think about how these attributes affect motion as they prepare for their trip to the playground.

Playground Pals

Introduce the Playground Pals activity to students by reminding them that they predicted, observed, and then explained the motion of their toys in the Physics in Motion Game Days activity. Tell students that they will use the same process as they investigate their bodies' forces on playground equipment. Introduce the playground equipment students will use during the activity (see Preparation for Lesson 1, p. 49), creating a POE chart with a row for each type of equipment. Tell students that they will first make predictions about what will happen when they apply different types of forces to the playground equipment. Hold a class discussion about each type of equipment and students' predictions. Document student ideas and findings on the POE chart throughout the activity, and have students record their predictions, observations, and explanations in STEM Research Notebook entries #7 and #8.

Ask students to share what they know about swings and how they move. Then ask:

- Will you swing higher when you are pushed by a friend or when you pump your own legs?

Ask students to share what they know about slides and how they work. Show students the squares of carpet, cardboard, and waxed paper, and then ask:

- Will you slide faster with nothing under you or with a piece of cardboard under you?

- Will you slide faster with nothing under you or with a piece of carpet (with the soft side against the slide) under you?

- Will you slide faster with nothing under you or with a piece of waxed paper under you?

Ask students to share what they know about seesaws and how they work. Then ask:

- What will happen when students of similar size sit at the same spot on opposite ends of the seesaw?

- What will happen when students of different sizes sit at the same spot on opposite ends of the seesaw?

STEM Research Notebook Entry #7

Before students engage with playground equipment, have them record in their STEM Research Notebooks their predictions about what will happen when they explore the playground equipment.

Students should work in pairs for this activity (for the seesaw activity, students should switch partners so that they seesaw with one similar-sized partners and one larger or smaller partner). Review playground safety guidelines with students beforehand.

STEM Research Notebook Entry #8

On a playground, have students engage in the following activities and document their observations in their STEM Research Notebooks:

Swing (observe which way they swing higher):

- Swing with their partner pushing them three times
- Swing by pumping their legs three times

Slide (observe which way they slide fastest):

- Slide down the slide with nothing under them
- Slide down the slide on a carpet square, with the soft side down
- Slide down the slide on a piece of cardboard
- Slide down the slide on a piece of waxed paper

Seesaw (observe whether one side of the seesaw moves up and the other down):

- With students of similar size sitting at the same spot on opposite ends of the seesaw
- With students of different sizes sitting on opposite ends of the seesaw

Then, hold a class discussion about students' observations, adding to the POE chart. Hold a class discussion in which students share ideas about whether their predictions were right or not, and why they think this is so.

Remind students that the next step is to explain their observations. Work as a class to formulate explanations for students' observations on each piece of playground equipment, incorporating the concepts of pushes and pulls and gravity.

Assess student learning by having students write six words that describe how things move. Have them draw and label three pictures depicting motion. Students should use a minimum of three vocabulary words.

Social Studies Connection: Hold a class discussion about safety guidelines for playgrounds, asking students to share the safety rules they know. Ask students to share their ideas about why these rules are in place, recording their responses on a class chart. Next, ask them to name some other activities for which there are safety rules (e.g., swimming, driving or riding in cars, riding bikes) and have them give examples of rules, recording students' responses. Discuss with students the role of rules and laws in keeping people safe and healthy. Conduct an interactive read-aloud of *I Can Be Safe: A First Look at Safety,* by Pat Thomas. After the read-aloud, ask students to share additional ideas about the reasons for safety rules and activities for which there are safety rules, adding their responses to the class list.

Evaluation/Assessment

Students may be assessed on the following performance tasks and other measures listed.

Performance Tasks

- Physics in Motion Game Days investigation

- Playground Pals investigation

- Lesson Assessment

Other Measures (using assessment rubric in Appendix B, p. 109)

- Teacher observations

- STEM Research Notebook entries

- Participation in teams during investigations

INTERNET RESOURCES

Formative assessment
- *http://stemteachingtools.org/pd/sessionc*

Isaac Newton
- *www.biography.com/news/how-isaac-newton-changed-our-world*

- *www.ducksters.com/biography/scientists/isaac_newton.php*

Forces
- *www.physics4kids.com/files/motion_intro.html*

- *www.physicsclassroom.com/class/newtlaws/Lesson-2/Types-of-Forces*

Bureau of Labor Statistics' *Occupational Outlook Handbook*
- *www.bls.gov/ooh/home.html*

KLEWS chart
- *www.nsta.org/publications/news/story.aspx?id=51519*

- *www.nsta.org/store/product_detail.aspx?id=10.2505/4/sc15_052_06_66*

Interactive read-alouds
- *www.readingrockets.org/article/repeated-interactive-read-alouds-preschool-and-kindergarten*

- *www.k5chalkbox.com/interactive-read-aloud.html*

- *www.readwritethink.org/professional-development/strategy-guides/teacher-read-aloud-that-30799.html*

Predict, Observe, Explain process
- *https://arbs.nzcer.org.nz/predict-observe-explain-poe*

"Sid the Science Kid: 'Sid's Super Kick,' part 2" video
- *www.dailymotion.com/video/x15oaoe*

Lesson Plan 2: Roller Coaster Fun!

In this lesson, students explore the concept of physics in motion by exploring roller coasters and working in teams to create virtual roller coasters.

ESSENTIAL QUESTIONS

- What kind of motion do riders experience on roller coasters?
- How do roller coasters move?

ESTABLISHED GOALS AND OBJECTIVES

At the conclusion of this lesson, students will be able to do the following:

- Describe how a roller coaster works
- Identify and explain gravity as a force that works on roller coaster cars
- Identify what safety precautions might be important for roller coaster riders
- Use the EDP to create and evaluate virtual roller coaster tracks
- Communicate and present findings about virtual roller coaster tracks
- Use technology to gather research information and communicate
- Identify several environmental impacts of roller coasters built within amusement parks
- Identify careers associated with roller coasters

TIME REQUIRED

- 6 days (approximately 30 minutes each day; see Tables 3.8–3.9, p. 36)

MATERIALS

Required Materials for Lesson 2

- STEM Research Notebooks
- Computers (1 per team) and internet access
- Books
  - *Archibald Frisby,* by Michael Chesworth (Farrar, Straus and Giroux, 1994)
  - *Gravity in Action: Roller Coasters!* by Joan Newton (PowerKids Press, 2009)

- *Roller Coaster,* by Marla Frazee (Houghton Mifflin Harcourt, 2006)
- Chart paper
- Markers

SAFETY NOTE

- Direct supervision is required during all aspects of this activity to make sure safety behaviors are followed and enforced.

CONTENT STANDARDS AND KEY VOCABULARY

Table 4.4 lists the content standards from the *NGSS, CCSS,* NAEYC, and the Framework for 21st Century Learning that this lesson addresses, and Table 4.5 (p. 64) presents the key vocabulary. Vocabulary terms are provided for both teacher and student use. Teachers may choose to introduce some or all of the terms to students.

Table 4.4. Content Standards Addressed in STEM Road Map Module Lesson 2

NEXT GENERATION SCIENCE STANDARDS

PERFORMANCE EXPECTATIONS

- K-PS2-1. Plan and conduct an investigation to compare the effects of different strengths or different directions of pushes and pulls on the motion of an object.

- K-PS2-2. Analyze data to determine if a design solution works as intended to change the speed or direction of an object with a push or a pull.

SCIENCE AND ENGINEERING PRACTICE

Planning and Carrying Out Investigations

- With guidance, plan and conduct an investigation in collaboration with peers.

DISCIPLINARY CORE IDEAS

PS2.A. Forces and Motion

- Pushes and pulls can have different strengths and directions.

- Pushing or pulling on an object can change the speed or direction of its motion and can start or stop it.

PS2.B. Types of Interactions

- When objects touch or collide, they push on one another and can change motion.

PS3.C. Relationship Between Energy and Forces

- A bigger push or pull makes things speed up or slow down more quickly.

Continued

Table 4.4. (*continued*)

CROSSCUTTING CONCEPT

Cause and Effect

- Simple tests can be designed to gather evidence to support or refute student ideas about causes.

COMMON CORE STATE STANDARDS FOR MATHEMATICS

MATHEMATICAL PRACTICES

- MP1. Make sense of problems and persevere in solving them.
- MP2. Reason abstractly and quantitatively.
- MP3. Construct viable arguments and critique the reasoning of others.
- MP4. Model with mathematics.
- MP5. Use appropriate tools strategically.
- MP6. Attend to precision.

MATHEMATICAL CONTENT

- K.MD.B.3. Classify objects into given categories; count the numbers of objects in each category and sort the categories by count.
- K.CC.C.6. Identify whether the number of objects in one group is greater than, less than, or equal to the number of objects in another group, e.g., by using matching and counting strategies.
- K.CC.C.7. Compare two numbers between 1 and 10 presented as written numerals.
- K.CC.B.4. Understand the relationship between numbers and quantities; connect counting to cardinality.
- K.CC.B.4a. When counting objects, say the number names in the standard order, pairing each object with one and only one number name and each number name with one and only one object.
- K.CC.B.4b. Understand that the last number name said tells the number of objects counted. The number of objects is the same regardless of their arrangement or the order in which they were counted.
- K.CC.B.4c. Understand that each successive number name refers to a quantity that is one larger.
- K.MD.A.1. Describe measurable attributes of objects, such as length or weight. Describe several measurable attributes of a single object.
- K.MD.A.2. Directly compare two objects with a measurable attribute in common, to see which object has "more of"/"less of" the attribute, and describe the difference. For example, directly compare the heights of two children and describe one child as taller/shorter.

Continued

Table 4.4. (*continued*)

> *COMMON CORE STATE STANDARDS FOR ENGLISH LANGUAGE ARTS*
>
> **READING STANDARDS**
> - RI.K.1. With prompting and support, ask and answer questions about key details in a text.
> - RI.K.3. With prompting and support, describe the connection between two individuals, events, ideas, or pieces of information in a text.
>
> **WRITING STANDARDS**
> - W.K.2. Use a combination of drawing, dictating, and writing to compose informative/ explanatory texts in which they name what they are writing about and supply some information about the topic.
> - W.K.5. With guidance and support from adults, respond to questions and suggestions from peers and add details to strengthen writing as needed.
> - W.K.7. Participate in shared research and writing projects (e.g., explore a number of books by a favorite author and express opinions about them).
>
> **SPEAKING AND LISTENING STANDARDS**
> - SL.K.1. Participate in collaborative conversations with diverse partners about *kindergarten topics and texts* with peers and adults in small and larger groups.
> - SL.K.3. Ask and answer questions in order to seek help, get information, or clarify something that is not understood.
> - SL.K.5. Add drawings or other visual displays to descriptions as desired to provide additional detail.
>
> **NATIONAL ASSOCIATION FOR THE EDUCATION OF YOUNG CHILDREN STANDARDS**
> - 2.G.02. Children are provided varied opportunities and materials to learn key content and principles of science.
> - 2.G.03. Children are provided with varied opportunities and materials that encourage them to use the five senses to observe, explore, and experiment with scientific phenomena.
> - 2.G.04. Children are provided with varied opportunities to use simple tools to observe objects and scientific phenomena.
> - 2.G.05. Children are provided with varied opportunities and materials to collect data and to represent and document their findings (e.g., through drawing or graphing).
> - 2.G.06. Children are provided with varied opportunities and materials that encourage them to think, question, and reason about observed and inferred phenomena.
> - 2.G.07. Children are provided with varied opportunities and materials that encourage them to discuss scientific concepts in everyday conversation.
> - 2.G.08. Children are provided with varied opportunities and materials that help them learn and use scientific terminology and vocabulary associated with the content areas.

Continued

Table 4.4. (*continued*)

> • 2.H.03. Technology is used to extend learning within the classroom and integrate and enrich the curriculum.
>
> **FRAMEWORK FOR 21ST CENTURY LEARNING**
> • Interdisciplinary Themes; Learning and Innovation Skills; Information, Media, and Technology Skills; Life and Career Skills

Table 4.5. Key Vocabulary for Lesson 2

Key Vocabulary	Definition
environment	the natural world around us, including both living things and nonliving things
friction	the force that causes a moving object to slow down when it is touching another object
habitat	an animal's or plant's home
inertia	the way an object at rest stays at rest or a moving object keeps moving until acted on by a force
roller coaster	amusement park ride that carries riders up and down hills and around curves in small open cars

TEACHER BACKGROUND INFORMATION
Roller Coasters

Students will begin to explore roller coasters in this lesson. Kindergartners may have limited experience with amusement parks and roller coasters. Although there are many kinds of roller coasters made from various materials (e.g., wood or metal) and with different types of rider seats (e.g., traditional cars on a track or individual seats suspended from overhead), the concept of roller coasters stressed in this lesson is that they are amusement park rides that carry riders up and down hills and around curves in small open cars. The focus of the lesson is on forces that act on roller coasters. For example, a pull is needed to move a roller coaster car up a hill, and gravity causes the cars to descend hills.

Students will also consider safety for roller coaster riders. Roller coasters typically employ either lap bars or shoulder harnesses (or both) to keep riders safe. Each roller coaster ride has its own posted set of rider requirements that typically encompass height and weight, as well as health. People with certain medical issues or conditions, such as neck injuries, heart problems, or pregnancy, are advised not to ride roller coasters. Roller coasters incorporate many safety features in their design; however, it is important that riders follow posted rules in order to ride safely. The following links provide relevant information on roller coaster safety:

- *www.wikihow.com/Ride-a-Roller-Coaster*

- *http://themeparks.lovetoknow.com/Roller_Coaster_Safety_Tips*

As part of the social studies connection for this lesson, students will consider environmental effects of roller coasters. These effects include the possibility of noise pollution that affects people who live nearby, movement (vibrations) in the ground around the roller coaster that could affect animals living nearby, and the use of trees for lumber to build wooden roller coasters. Environmental effects of amusement parks more broadly include the clearing of large amounts of land and potentially harming animal habitats, noise pollution, and the waste created by visitors to amusement parks. The following link provides information about the impacts amusement parks can have on the environment: *www.bizfluent.com/info-8483212-environmental-come-making-theme-park.html*.

Inertia and Friction

Students will experiment with an online roller coaster building program in this lesson and will experience the impact of inertia and friction on the virtual roller coasters. Newton's first law of motion relates to inertia and states that an object at rest stays at rest until acted on by a force and that an object that is moving will continue moving in the same direction until acted on by a force. Students will observe that their virtual roller coasters stop moving at some point. This cessation of motion is due to inertia caused by the friction of the car against a track. Kindergarten students can observe friction as a force that prevents sliding or rolling objects from moving indefinitely. More information about Newton's first law of motion can be found at *www.ducksters.com/science/laws_of_motion.php*. More information about friction can be found at *www.ducksters.com/science/friction.php*.

Engineering

Students begin to gain an understanding of engineering as a profession in this lesson as they learn to use the EDP to design roller coasters. Students should understand that engineers are people who design and build products and systems in response to human needs. For an overview of the various types of engineering professions, see the following websites:

- *www.engineergirl.org/33/TryOnACareer*

- *www.nacme.org/types-of-engineering*

- *www.sciencekids.co.nz/sciencefacts/engineering/typesofengineeringjobs.html*

Engineering Design Process

Students should understand that engineers need to work in groups to accomplish their work and that collaboration is important for designing solutions to problems. Students will use the EDP, the same process that professional engineers use in their work, in this lesson. A graphic representation of the EDP is provided at the end of this lesson (p. 74). You may wish to provide each student with a copy of the EDP graphic or enlarge it and post it in a prominent place in your classroom for student reference throughout the module. Be prepared to review each step of the EDP with students, and emphasize that the process is not a linear one—at any point in the process, they may need to return to a previous step. The steps of the process are as follows:

1. *Define.* Describe the problem you are trying to solve, identify what materials you are able to use, and decide how much time and help you have to solve the problem.

2. *Learn.* Brainstorm solutions and conduct research to learn about the problem you are trying to solve.

3. *Plan.* Plan your work, including making sketches and dividing tasks among team members if necessary.

4. *Try.* Build a device, create a system, or complete a product.

5. *Test.* Now, test your solution. This might be done by conducting a performance test, if you have created a device to accomplish a task, or by asking for feedback from others about their solutions to the same problem.

6. *Decide.* Based on what you found out during the Test step, you can adjust your solution or make changes to your device.

After completing all six steps, students can share their solution or device with others. This represents an additional opportunity to receive feedback and make modifications based on that feedback. The following are additional resources about the EDP:

- *www.sciencebuddies.org/engineering-design-process/engineering-design-compare-scientific-method.shtml*

- *www.pbslearningmedia.org/resource/phy03.sci.engin.design.desprocess/what-is-the-design-process*

COMMON MISCONCEPTIONS

Students will have various types of prior knowledge about the concepts introduced in this lesson. Table 4.6 outlines some common misconceptions students may have concerning these concepts. Because of the breadth of students' experiences, it is not possible to anticipate every misconception that students may bring as they approach this lesson. Incorrect or inaccurate prior understanding of concepts can influence student learning in the future, however, so it is important to be alert to misconceptions such as those presented in the table.

Table 4.6. Common Misconceptions About the Concepts in Lesson 2

Topic	Student Misconception	Explanation
Engineers and the engineering design process	All engineers are people who drive trains.	While people who drive trains may be called *engineers*, other engineers are people who use science, technology, and mathematics to build machines, products, and structures that meet people's needs.
	Engineers use only science and mathematics to do their work.	Engineers often use science and mathematics in their work, but also use other knowledge to solve problems. Engineers need to understand many other things to design products, like how people use products, what people's needs are, and how the natural environment affects materials.
	Engineers work alone to build things.	Engineers often work in teams and use a process to solve problems. The process involves creative thinking, research, and planning in addition to building and testing products.

PREPARATION FOR LESSON 2

Review the Teacher Background Information section (p. 64), assemble the materials for the lesson, duplicate the EDP graphic (p. 74) if you wish to hand it out to students or enlarge it to post in the classroom, and preview the video recommended in the Learning Components section that follows.

You should identify and explore a simple virtual roller coaster building website such as *http://static.lawrencehallofscience.org/kidsite/portfolio/rollercoaster-builder* before the start

of the lesson. Be prepared to provide students with instructions about how to use the site.

If you choose to incorporate the optional field trip to a local amusement park (see p. 72), make appropriate preparations for this trip.

LEARNING COMPONENTS
Introductory Activity/Engagement

Connection to the Challenge: Begin each day of this lesson by directing students' attention to the module challenge, the Roller Coaster Design Challenge. Hold a brief class discussion each day about how students' learning in the previous days' lessons contributed to their ability to complete the challenge. You may wish to create a class list of key ideas on chart paper.

Science Class and ELA Connection: Hold a class discussion about roller coasters to access students' prior knowledge, ideas, and opinions about roller coasters. Track student responses on a KLEWS chart. Following agreed-upon rules for discussions, ask your students:

- What do you know about roller coasters?

- What is a roller coaster?

- Who has ridden on a roller coaster? When? Where?

- What do roller coasters look like?

- How did you feel when you rode the roller coaster?

- What did you like?

- What did you not like?

- Did you have to be a certain height to ride the roller coaster?

- What do you wonder about roller coasters?

Show students a video of a simple roller coaster (preferably a video that provides the perspective of a roller coaster rider), such as the "Small Coaster Wild Mouse Roller Coaster Front Seat POV Teine Olympia Japan" video at *www.youtube.com/watch?v=ooVHYjvg4Vk*. After watching the video, ask students how the "virtual" experience of riding a roller coaster made them feel and what they noticed about the roller coaster or their experience. Add students' ideas to the KLEWS chart.

Next, explore roller coasters as a class by conducting an interactive read-aloud of the book *Roller Coaster*, by Marla Frazee. After the read-aloud, ask students what they learned about roller coasters, adding their ideas to the KLEWS chart.

STEM Research Notebook Entry #9

Have students draw a picture of a roller coaster in their STEM Research Notebooks.

Mathematics Connection: Refer to the KLEWS chart and point out descriptions of roller coasters students shared (e.g., fast, high). Encourage students to compare these descriptions with other things they are familiar with by asking them to fill in the blanks in statements with comparing words such as the following:

- A roller coaster is _____ (faster or slower) than a car.

- A roller coaster is _____ (higher or lower) than the ceiling of our classroom.

- A roller coaster car is _____ (bigger or smaller) than a car that drives on the road.

Social Studies Connection: Encourage students to connect roller coasters with careers by holding a class discussion about the following questions, listing students' answers on chart paper:

- How do you think roller coasters are made?

- Who makes roller coasters?

Next, explore innovation and roller coaster design as a class by conducting an interactive read-aloud of *Archibald Frisby,* by Michael Chesworth, and then watching the video about students designing a roller coaster at *www.pbslearningmedia.org/resource/phy03.sci. phys.mfe.zcoaster/designing-a-roller-coaster*.

STEM Research Notebook Entry #10

After the interactive read-aloud and watching the video, have students document what they learned about roller coaster design in their STEM Research Notebooks, using both words and pictures.

Activity/Exploration

Science Class and ELA, Mathematics, and Social Studies Connections: Introduce students to the concept that engineers are people who design and build things like roller coasters and that they work in teams and use the engineering design process, called the EDP for short, to do this. Introduce the steps of the EDP to students, and tell them that they will use the EDP to construct roller coasters using a computer program in the Design Time! activity.

Design Time!

Group students in teams of three or four for this activity. Students will remain in these teams throughout the remainder of the module. Guide teams through each step of the EDP as they prepare for the activity and design their virtual roller coasters. Track student responses for each step of the EDP on chart paper. For the Learn step of the EDP, you may wish to review the instructions on the website with students. For the Plan step, you might have students plan how they will share the responsibilities of building the coaster (how will team members take turns as they design the roller coaster?). Then, model use of the program for the class and give students the opportunity to work with the coaster design website.

Remind students that engineers keep track of their ideas, observations, and explanations in research notebooks. Tell students that they are acting as engineers in this activity and will record their ideas, observations, and explanations in their STEM Research Notebooks.

STEM Research Notebook Entry #11

Have students draw their team's best roller coaster design in their STEM Research Notebooks.

Then, have students share their team's most successful virtual roller coaster designs with the class, by showing either their designs on the computer or their sketches in their STEM Research Notebooks. Have students explain their designs.

Explanation

Science Class and ELA and Mathematics Connections: Have students share their observations about their online roller coaster designs, asking them questions such as the following:

- What were some of the differences between your best roller coaster and the others you designed that did not work as well?

- Did your car stop before it reached the end of the track in some of your designs? Why do you think this is so?

- Did your car ever stop when it was going downhill?

- Did your car ever stop when it was going uphill?

- Did your car ever stop when it was on a flat section of the track?

- On what section of the track did your car move the fastest? The slowest?

Record students' responses on chart paper. Remind them about their discussion of gravity in Lesson 1. Ask students if they observed gravity at work with their online roller coasters (e.g., the car never stopped while it was going downhill because gravity was pulling it down). Next, ask why they think their cars stopped when they did. Introduce the concept of inertia, and ask students if they saw anything putting a force on the car that would make it stop. Ask what happens when they slide on a slippery surface—do they keep going forever or do they stop? Introduce the concept of friction as a force that a surface puts on a moving object that makes it slow down or stop.

Conduct an interactive read-aloud of the book *Gravity in Action: Roller Coasters!* by Joan Newton to support students' understanding of science concepts associated with roller coaster design and function. After the read-aloud, hold a class discussion about how gravity and other forces influence the motion of a roller coaster car.

STEM Research Notebook Entry #12

Have students document what they learned about gravity and other science concepts associated with roller coasters in their STEM Research Notebooks.

Social Studies Connection: Remind students of the discussion about safety rules from Lesson 1. Ask students if they think there are any safety rules for roller coasters. Explain that roller coasters are built with safety features such as lap bars or shoulder harnesses and that signs posted at the entrance of roller coasters provide important safety information for riders. Ask students to give some examples of safety guidelines for roller coasters, creating a class list of students' responses (e.g., fasten the shoulder harness securely, keep hands and feet inside the car when it is moving, don't try to stand up when the roller coaster is moving, make sure the riders are big enough for the lap bar or shoulder harness to fit securely).

Elaboration/Application of Knowledge

Science Class and ELA and Social Studies Connections: Hold a class discussion about the environment around roller coasters. Remind students that roller coasters are typically found in amusement parks that have many visitors, sidewalks, and parking lots. Ask students to share their ideas of ways that roller coasters and amusement parks might change their surroundings. Show the pictures of the open field and the amusement park (Figures 4.2 and 4.3 on p. 75). Prepare a two-column table on chart paper, with one column labeled "Field" and the other "Amusement Park." Ask students to describe the field and what they might find there (e.g., plants, animals, nature) and then to describe the amusement park and what they might find there (e.g., many rides, food, sidewalks and paved areas, lights, people), recording their descriptive words on the chart. Next, ask students to name some differences between these environments.

Ask students to consider what would change about the field if an amusement park were built there. Ask what they think would happen to the plants and the homes of animals living in the field. Then, ask them to consider how people living near the field might be affected after the amusement park was built. Ask whether they think it is important to think carefully about where an amusement park should be built. Why?

If you decided to incorporate a field trip to an amusement park into the lesson, try to secure a picture of what the area looked like before the amusement park was built to show students. During the visit, have students observe safety features of roller coasters and make observations about the impacts of roller coasters and amusement parks in general on the natural environment.

Mathematics Connection: Introduce the idea of speed to students as a measure of how fast something goes. Ask students to compare the speeds of sets of items such as a turtle and a rabbit, a car and a person walking, a roller coaster going down a hill and a person running. Tell students that speed is a measure of the distance a thing moves in a certain period of time. As an example, have the class find the speed of a student walking. To do this, choose one student to walk across the room. When you say, "Start," have the student begin walking at an ordinary pace and counting his or her steps while you time 5 seconds on a clock or a stopwatch. Write the student's speed on the board or on chart paper using the format "___ steps in 5 seconds." Repeat the process, this time having the student walk as fast as he or she can across the room, counting steps for 5 seconds. Write the student's speed using the same format as above. Have students compare the number of steps and identify one as faster and one as slower.

Evaluation/Assessment

Students may be assessed on the following performance tasks and other measures listed.

Performance Tasks

- Virtual roller coaster tracks

Other Measures (using assessment rubric in Appendix B, p. 109)

- Teacher observations

- STEM Research Notebook entries

- Participation in teams during investigations

- Participation in class discussions

4

INTERNET RESOURCES

Roller coaster safety
- *www.wikihow.com/Ride-a-Roller-Coaster*

- *http://themeparks.lovetoknow.com/Roller_Coaster_Safety_Tips*

Amusement parks and the environment
- *www.bizfluent.com/info-8483212-environmental-come-making-theme-park.html*

Newton's first law of motion
- *www.ducksters.com/science/laws_of_motion.php*

Friction
- *www.ducksters.com/science/friction.php*

Engineering careers
- *www.engineergirl.org/33/TryOnACareer*

- *www.nacme.org/types-of-engineering*

- *www.sciencekids.co.nz/sciencefacts/engineering/typesofengineeringjobs.html*

EDP resources
- *www.sciencebuddies.org/engineering-design-process/engineering-design-compare-scientific-method.shtml*

- *www.pbslearningmedia.org/resource/phy03.sci.engin.design.desprocess/what-is-the-design-process*

Virtual roller coaster building website
- *http://static.lawrencehallofscience.org/kidsite/portfolio/rollercoaster-builder*

"Small Coaster Wild Mouse Roller Coaster Front Seat POV Teine Olympia Japan" video
- *www.youtube.com/watch?v=ooVHYjvg4Vk*

"Designing a Roller Coaster" video
- *www.pbslearningmedia.org/resource/phy03.sci.phys.mfe.zcoaster/designing-a-roller-coaster*

ENGINEERING DESIGN PROCESS

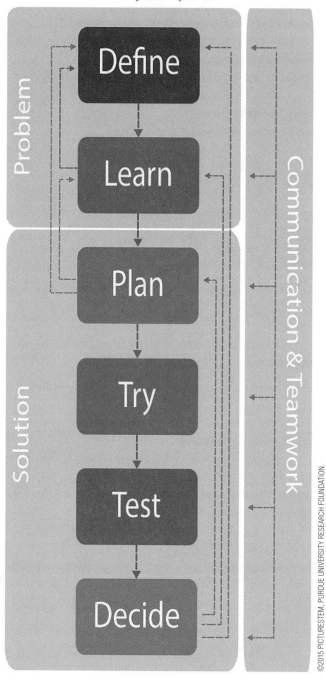

Figure 4.2. Open Field

Figure 4.3. Amusement Park

Note: Full-color versions of these images are available on the book's Extras page at *www.nsta.org/roadmap-physics.*

Lesson Plan 3: Roller Coaster Design Challenge

In this lesson, students work in teams to use the EDP and apply their learning from previous lessons to design, build, and test various marble track roller coasters.

ESSENTIAL QUESTIONS

- How can we use the EDP to create a marble track roller coaster?

- On what kind of track does a marble roll the fastest?

ESTABLISHED GOALS AND OBJECTIVES

At the conclusion of this lesson, students will be able to do the following:

- Use the EDP to create marble track roller coasters

- Communicate and present findings about their marble track roller coasters

- Use their understanding of safety to create safety guidelines for their roller coasters

- Use their understanding of roller coasters to create flyers about their roller coasters

TIME REQUIRED

- 5 days (approximately 30 minutes each day; see Table 3.10, p. 37)

MATERIALS

Required Materials for Lesson 3

- STEM Research Notebooks

- Computer with internet access

- Book

  - *Gravity in Action: Roller Coasters!* by Joan Newton (PowerKids Press, 2009)

- Chart paper

- Markers

- Tape measure

- Stopwatch

- Safety glasses with side shields or safety goggles (1 pair per student)

Additional Materials for Roller Coaster Design Challenge (per team of 3–4 students unless otherwise noted)

- 2 pieces of chart paper

- 1 marble

- Ruler

- Materials to elevate ramps (e.g., plastic shoe boxes or stacks of books)

- 2 cardboard paper towel tubes or 1 cardboard wrapping paper tube, cut in half lengthwise to create U-shaped marble tracks

- 1½ inch foam pipe insulation (two 4-foot sections and two 2-foot sections), cut in half lengthwise to create U-shaped marble tracks

- Masking tape

- Poster board

- Markers

- White paper (1 sheet per student)

- Crayons (1 set per student)

SAFETY NOTES

1. All students must wear safety glasses with side shields or goggles during all phases of this inquiry activity.

2. Direct supervision is required during all aspects of this activity to make sure safety behaviors are followed and enforced.

3. Make sure any items dropped on the floor or ground are picked up to avoid slip- or trip-and-fall hazards.

4. Make sure all fragile materials, furniture, and equipment are out of the activity area before students begin construction and operation of the roller coaster.

5. Have students wash hands with soap and water after completing the activity.

CONTENT STANDARDS AND KEY VOCABULARY

Table 4.7 (p. 78) lists the content standards from the *NGSS*, *CCSS*, NAEYC, and the Framework for 21st Century Learning that this lesson addresses, and Table 4.8 (p. 80) presents the key vocabulary. Vocabulary terms are provided for both teacher and student use. Teachers may choose to introduce some or all of the terms to students.

Table 4.7. Content Standards Addressed in STEM Road Map Module
Lesson 3

NEXT GENERATION SCIENCE STANDARDS

PERFORMANCE EXPECTATIONS

- K-PS2-1. Plan and conduct an investigation to compare the effects of different strengths or different directions of pushes and pulls on the motion of an object.

- K-PS2-2. Analyze data to determine if a design solution works as intended to change the speed or direction of an object with a push or a pull.

SCIENCE AND ENGINEERING PRACTICE

Planning and Carrying Out Investigations

- With guidance, plan and conduct an investigation in collaboration with peers.

DISCIPLINARY CORE IDEAS

PS2.A. Forces and Motion

- Pushes and pulls can have different strengths and directions.

- Pushing or pulling on an object can change the speed or direction of its motion and can start or stop it.

PS2.B. Types of Interactions

- When objects touch or collide, they push on one another and can change motion.

PS3.C. Relationship Between Energy and Forces

- A bigger push or pull makes things speed up or slow down more quickly.

CROSSCUTTING CONCEPT

Cause and Effect

- Simple tests can be designed to gather evidence to support or refute student ideas about causes.

COMMON CORE STATE STANDARDS FOR MATHEMATICS

MATHEMATICAL PRACTICES

- MP1. Make sense of problems and persevere in solving them.

- MP2. Reason abstractly and quantitatively.

- MP3. Construct viable arguments and critique the reasoning of others.

- MP4. Model with mathematics.

- MP5. Use appropriate tools strategically.

- MP6. Attend to precision.

Continued

Table 4.7. (*continued*)

MATHEMATICAL CONTENT

- K.CC.C.7. Compare two numbers between 1 and 10 presented as written numerals.

- K.CC.B.4. Understand the relationship between numbers and quantities; connect counting to cardinality.

- K.CC.B.4a. When counting objects, say the number names in the standard order, pairing each object with one and only one number name and each number name with one and only one object.

- K.CC.B.4b. Understand that the last number name said tells the number of objects counted. The number of objects is the same regardless of their arrangement or the order in which they were counted.

- K.CC.B.4c. Understand that each successive number name refers to a quantity that is one larger.

- K.MD.A.1. Describe measurable attributes of objects, such as length or weight. Describe several measurable attributes of a single object.

- K.MD.A.2. Directly compare two objects with a measurable attribute in common, to see which object has "more of"/"less of" the attribute, and describe the difference. For example, directly compare the heights of two children and describe one child as taller/shorter.

COMMON CORE STATE STANDARDS FOR ENGLISH LANGUAGE ARTS

READING STANDARD

- RI.K.1. With prompting and support, ask and answer questions about key details in a text.

WRITING STANDARD

- W.K.2. Use a combination of drawing, dictating, and writing to compose informative/explanatory texts in which they name what they are writing about and supply some information about the topic.

SPEAKING AND LISTENING STANDARDS

- SL.K.1. Participate in collaborative conversations with diverse partners about *kindergarten topics and texts* with peers and adults in small and larger groups.

- SL.K.3. Ask and answer questions in order to seek help, get information, or clarify something that is not understood.

- SL.K.5. Add drawings or other visual displays to descriptions as desired to provide additional detail.

Continued

Table 4.7. (*continued*)

> **NATIONAL ASSOCIATION FOR THE EDUCATION OF YOUNG CHILDREN STANDARDS**
> - 2.G.02. Children are provided varied opportunities and materials to learn key content and principles of science.
> - 2.G.03. Children are provided with varied opportunities and materials that encourage them to use the five senses to observe, explore, and experiment with scientific phenomena.
> - 2.G.04. Children are provided with varied opportunities to use simple tools to observe objects and scientific phenomena.
> - 2.G.05. Children are provided with varied opportunities and materials to collect data and to represent and document their findings (e.g., through drawing or graphing).
> - 2.G.06. Children are provided with varied opportunities and materials that encourage them to think, question, and reason about observed and inferred phenomena.
> - 2.G.07. Children are provided with varied opportunities and materials that encourage them to discuss scientific concepts in everyday conversation.
> - 2.G.08. Children are provided with varied opportunities and materials that help them learn and use scientific terminology and vocabulary associated with the content areas.
> - 2.H.03. Technology is used to extend learning within the classroom and integrate and enrich the curriculum.
>
> **FRAMEWORK FOR 21ST CENTURY LEARNING**
> - Interdisciplinary Themes; Learning and Innovation Skills; Information, Media, and Technology Skills; Life and Career Skills

Table 4.8. Key Vocabulary for Lesson 3

Key Vocabulary	Definition
failure	when something does not work as planned and needs to be changed or improved
improvement	something we do to make a thing work better
maximum	the largest amount allowed or possible
requirement	something that must be done
success	when something works as planned
tunnel	a path that is covered on the sides and open at the ends so that an object or person can move through it

TEACHER BACKGROUND INFORMATION

In this lesson, students work in teams to use the EDP to create marble track roller coasters. The following links provide examples of creating marble track roller coasters and additional information that may be useful to you as you guide students' work:

- *www.buggyandbuddy.com/science-kids-create-marble-run*

- *www.exploratorium.edu/tinkering/projects/marble-machines*

COMMON MISCONCEPTIONS

Students will have various types of prior knowledge about the concepts introduced in this lesson. Table 4.9 outlines a common misconception students may have concerning these concepts. Because of the breadth of students' experiences, it is not possible to anticipate every misconception that students may bring as they approach this lesson. Incorrect or inaccurate prior understanding of concepts can influence student learning in the future, however, so it is important to be alert to misconceptions such as the one presented in the table.

Table 4.9. Common Misconception About the Concepts in Lesson 3

Topic	Student Misconception	Explanation
Engineering design process	Engineers' most important job is to build things.	While engineers do build machines, products, and structures, they must first go through a careful process of researching and planning before they begin to build. By doing this, they can be sure that the products they create meet people's needs in the best way possible. In fact, after engineers have completed building a machine or product, they often test it and go back to the research and planning stages to improve their designs.

PREPARATION FOR LESSON 3

Review the Teacher Background Information section (p. 81), assemble the materials for the lesson, and duplicate the student handouts.

This lesson involves students working as teams to address the module challenge. It may be useful to have one adult or older student per team to provide individualized support to facilitate and guide students' work. You should ensure that classroom helpers understand the EDP as a way to structure students' teamwork. Label pieces of chart paper with one step of the EDP (Define, Learn, Plan, Try, Test, Decide) on each to guide teams' work and record students' thoughts and ideas.

Prepare a work station for each team with a set of challenge materials and sufficient space for the team to build a marble track that is 8 feet long. Use masking tape to mark a starting point on the floor for each team. Place another piece of masking tape on the floor 8 feet from each starting point to mark the maximum length of the roller coasters. Prepare a class list with the following requirements for the marble track roller coasters, including simple diagrams as appropriate:

- The roller coaster must have both downhill and uphill sections.

- The roller coaster must have a flat section that is at least as long as a ruler.

- The roller coaster must have a tunnel (a covered section).

- The track can be no longer than 8 feet.

- Teams may use only the materials provided, but they do not need to use all the materials.

- The marble must be able to roll without stopping from the starting point to the other end of the track.

- The marble must be able to roll from one end of the track to the other without falling off.

- Teams must use the EDP to design and build their roller coasters.

LEARNING COMPONENTS
Introductory Activity/Engagement

Connection to the Challenge: Begin each day of this lesson by directing students' attention to the module challenge, the Roller Coaster Design Challenge. Hold a brief class discussion each day about how students' learning in the previous days' lessons contributed to their ability to complete the challenge. You may wish to create a class list of key ideas on chart paper.

Science Class: Hold a class discussion about how using the EDP worked when teams designed their virtual roller coasters in Lesson 2. Tell students that they will use the EDP in their challenge in this lesson. Remind students of the module challenge, and tell them that they are going to act as engineers in this lesson as they create and test various marble tracks. Review the set of materials that each team will have to use, showing students examples of each material. Next, tell students that their roller coasters will have to meet a set of requirements. Show students the list of requirements on chart paper that you prepared and review the list with the class.

ELA Connection: Revisit relevant content from the book *Gravity in Action: Roller Coasters!* by Joan Newton. Ask students how they can use their learning from this book as they design their roller coasters. Encourage them to use the science vocabulary they have learned throughout the module in their discussion, and document student ideas on chart paper.

Mathematics Connection: Ask students to share their ideas about how large their roller coaster designs will be, asking them to consider the materials they have to work with. Measure a distance of 8 feet using a tape measure to show students. Tell students that their team's roller coaster can be as much as 8 feet long but no longer, and that it can be shorter than 8 feet as long as they include the required components.

Social Studies Connection: Remind students of their discussions about safety rules for roller coasters in Lesson 2. Ask students how they will know that their team's roller coaster is safe for the marble "rider" (the marble will not fall off the track).

Activity/Exploration

Science Class and Mathematics and ELA Connections: As a class, review the steps of the EDP. Then, group students into teams of 3 or 4, and tell them that each team will go through all the steps of the EDP as they design their marble tracks. Ask students to name the steps they will need to take before they build their tracks (Define, Learn, Plan).

Post the piece of chart paper labeled "Define." Ask students, "What is the problem your team needs to solve?" Record students' ideas on the chart paper, asking students to explain their responses (e.g., if students respond that the problem is how to create the best roller coaster, ask students to define "best"). Guide students to refine their understanding of the problem as how to build a marble track roller coaster that meets the requirements they were given.

Next, post the piece of chart paper labeled "Learn." Ask students what they need to know to solve the problem they identified (e.g., what the requirements are, what makes the marble move), recording students' ideas. Remind students of the science concepts they learned about in the previous lessons and in their activities, recording what they know that will be helpful to them as they design their roller coasters. The chart may include items such as the following:

- We know that gravity moves things from a higher place to a lower place.

- We know that once the marble starts moving, it will not stop unless another force acts on it.

- We know that the height of a hill affects how fast the marble moves.

After you have created the list, ask students for their ideas about how they can use this knowledge to create their roller coasters.

Next, post the piece of chart paper labeled "Plan." Tell students that this will be their chance to sketch a track that they think will work well. Remind students that their tracks must include uphill, downhill, and flat sections.

STEM Research Notebook Entry #13

Have students sketch their ideas for a roller coaster that meets the requirements in their STEM Research Notebooks.

After students have sketched their designs, have each team member share his or her design sketch with the team. Remind students that the roller coaster must meet the listed requirements, and the team must be able to build it using the materials provided. Have each team decide on one member's design to use. You might wish to allow teams to choose elements from various members' designs to create a final team design and then sketch that design on chart paper. Review each team's sketch and ensure that the team has identified what materials to use for each portion of its roller coaster.

Post the piece of chart paper labeled "Try." Tell students that they will now build their roller coasters, using the sketches their teams created to show them how it should look. After teams have had time to build their tracks, ask students for their ideas about how they will know if their design is a success (e.g., the marble keeps rolling to the end of the track). You may wish to make suggestions to students such as the following:

- Tape one end of the track to a desktop, and send the marble down to start it.

- Create hills by placing something under the tracks (e.g., books or shoebox) at various points.

- Add more hills.

Post the piece of chart paper labeled "Test" on the wall, and tell students that running the marble on the track and making sure it meets the requirements is their test. Ask students for their ideas about what they should do if the track does not work as they expected when they test it. Post the piece of chart paper labeled "Decide," and tell students that after they test their track designs, they will then decide whether their tracks are the best they can be and, if not, how they can improve them. Allow students time to build, test, and improve their tracks.

Social Studies Connection: Have students work in their teams to create a list of rules the team would post if its roller coaster were built in an amusement park. Have students document these rules using words and pictures on a piece of poster board.

Explanation

Science Class and Mathematics, ELA, and Social Studies Connections: After all teams have created what they think are the best tracks, have teams prepare to share their designs with the class by completing the next STEM Research Notebook entry.

STEM Research Notebook Entry #14

Have students draw their team's final track design in their STEM Research Notebooks and then document how they knew this was the best design and how they saw gravity at work in their roller coaster.

Next, have students share their marble track roller coasters and the safety rules they created. Students should demonstrate their roller coasters, explain how they knew this was the best design, and discuss how they used gravity. Hold a class discussion on the physics concepts (e.g., gravity, friction) that contributed to the success or failure of the marble tracks.

After each team has shared its design, use a stopwatch to time the marble on each track, creating a class list of the times. Ask students if they can tell from this list which track is the fastest. Remind students that speed is a measure of the distance covered over a certain amount of time, but you measured only time. Next, time the marbles from the starting points to the 8-foot tape marks. (For tracks that are less than 8 feet long, the marbles will have to roll off the track and continue to the tape mark.) Then, have the class compare speeds for each roller coaster.

Elaboration/Application of Knowledge

Science Class: After all teams have shared their track designs, allow teams to design additional marble tracks based on what they learned from designing their previous tracks and from observing other teams' tracks.

ELA, Mathematics, and Social Studies Connections: Have each team work together to plan a design for a theme park roller coaster. Have students sketch the design on chart paper, give the roller coaster a name, and use color or a theme to make the roller coaster appealing to riders. Next, help each student create a flyer on white paper for the team's amusement park roller coaster. The flyer should include the ride name, how gravity is used, why the rider should choose this roller coaster, and the time it will take a rider to ride the roller coaster.

Evaluation/Assessment

Students may be assessed on the following performance tasks and other measures listed.

Performance Tasks

- Construction of marble track roller coasters

- Presentation of marble track roller coasters

- Flyers for amusement park roller coasters

Other Measures (using assessment rubric in Appendix B, p. 109)

- Teacher observations

- STEM Research Notebook entries

- Participation in teams during investigations

INTERNET RESOURCES

Marble track roller coasters
- *www.buggyandbuddy.com/science-kids-create-marble-run*

- *www.exploratorium.edu/tinkering/projects/marble-machines*

SUGGESTED BOOKS

- *Eat My Dust! Henry Ford's First Race,* by Monica Kulling (Random House, 2004)

- *Forces Make Things Move,* by Kimberly Brubaker Bradley (HarperCollins, 2005)

- *How Things Move,* by Don L. Curry (Capstone Press, 2000)

- *Motion,* by Rebecca Olien (Capstone Press, 2004)

- *Motion: Push and Pull, Fast and Slow,* by Darlene Stille (Picture Window Books, 2004)

- *Move It! Motion, Forces, and You,* by Adrienne Mason (Kids Can Press, 2005)

- *Newton and Me,* by Lynne Mayer (Mayer Sylvan Dell, 2010)

- *Oscar and the Cricket: A Book About Moving and Rolling,* by Geoff Waring (Candlewick Press, 2008)

- *Push and Pull,* by Patricia J. Murphy (Children's Press, 2002)

- *Push and Pull,* by Lola M. Schaefer (Capstone Press, 2000)

- *Sheep in a Jeep,* by Nancy E. Shaw (Houghton Mifflin, 1986)

REFERENCE

Koehler, C., M. A. Bloom, and A. R. Milner. 2015. The STEM Road Map for grades K–2. In *STEM Road Map: A framework for integrated STEM education,* ed. C. C. Johnson, E. E. Peters-Burton, and T. J. Moore, 41–67. New York: Routledge. *www.routledge.com/products/9781138804234.*

TRANSFORMING LEARNING WITH PHYSICS IN MOTION AND THE *STEM ROAD MAP CURRICULUM SERIES*

Carla C. Johnson

This chapter serves as a conclusion to the Physics in Motion integrated STEM curriculum module, but it is just the beginning of the transformation of your classroom that is possible through use of the *STEM Road Map Curriculum Series.* In this book, many key resources have been provided to make learning meaningful for your students through integration of science, technology, engineering, and mathematics, as well as social studies and English language arts, into powerful problem- and project-based instruction. First, the Physics in Motion curriculum is grounded in the latest theory of learning for students in kindergarten specifically. Second, as your students work through this module, they engage in using the engineering design process (EDP) and build prototypes like engineers and STEM professionals in the real world. Third, students acquire important knowledge and skills grounded in national academic standards in mathematics, English language arts, science, and 21st century skills that will enable their learning to be deeper, retained longer, and applied throughout, illustrating the critical connections within and across disciplines. Finally, authentic formative assessments, including strategies for differentiation and addressing misconceptions, are embedded within the curriculum activities.

The Physics in Motion curriculum in the Cause and Effect STEM Road Map theme can be used in single-content classrooms (e.g., science) where there is only one teacher or expanded to include multiple teachers and content areas across classrooms. Through the exploration of the Roller Coaster Design Challenge, students engage in a real-world STEM problem on the first day of instruction and gather necessary knowledge and skills along the way in the context of solving the problem.

The other topics in the *STEM Road Map Curriculum Series* are designed in a similar manner, and NSTA Press has additional volumes in this series for this and other grade levels and plans to publish more. The volumes covering Innovation and Progress have been published and are as follows:

- *Amusement Park of the Future, Grade 6*

- *Construction Materials, Grade 11*

- *Harnessing Solar Energy, Grade 4*

- *Transportation in the Future, Grade 3*

- *Wind Energy, Grade 5*

The volumes covering The Represented World have also been published and are as follows:

- *Car Crashes, Grade 12*

- *Improving Bridge Design, Grade 8*

- *Investigating Environmental Changes, Grade 2*

- *Packaging Design, Grade 6*

- *Patterns and the Plant World, Grade 1*

- *Radioactivity, Grade 11*

- *Rainwater Analysis, Grade 5*

- *Swing Set Makeover, Grade 3*

The tentative list of other books includes the following themes and subjects:

- Cause and Effect
  - Earth on the move
  - Healthy living
  - Human impacts on our climate
  - Influence of waves
  - Natural hazards
- Sustainable Systems
  - Composting: Reduce, reuse, recycle
  - Creating global bonds

- Hydropower efficiency

- System interactions

- Optimizing the Human Experience

  - Genetically modified organisms

  - Mineral resources

  - Rebuilding the natural environment

  - Water conservation: Think global, act local

If you are interested in professional development opportunities focused on the STEM Road Map specifically or integrated STEM or STEM programs and schools overall, contact the lead editor of this project, Dr. Carla C. Johnson (*carlacjohnson@ncsu.edu*), associate dean and professor of science education and executive director of the William and Ida Friday Institute at North Carolina State University. Someone from the team will be in touch to design a program that will meet your individual, school, or district needs.

APPENDIX A

STEM RESEARCH NOTEBOOK TEMPLATES

MY STEM RESEARCH NOTEBOOK

PHYSICS IN MOTION

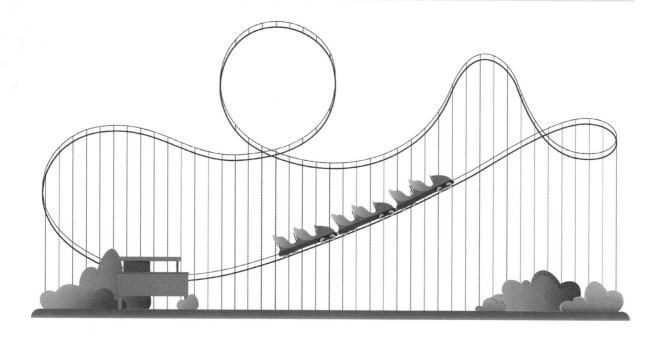

Name:

_____

- -

_____

Name: _____   Date: _____

STEM RESEARCH NOTEBOOK ENTRY #1 (LESSON PLAN 1)

Draw and label three things that you can move, and show how they move:

Name: _____   Date: _____

STEM RESEARCH NOTEBOOK ENTRY #2 (LESSON PLAN 1)

I learned ...

Name: _____   Date: _____

STEM RESEARCH NOTEBOOK ENTRY #3 (LESSON PLAN 1)

I learned ...

Name: _____

Date: _____

STEM RESEARCH NOTEBOOK ENTRY #4 (LESSON PLAN 1)

VOCABULARY WORDS

Vocabulary Word	Definition	Picture

Name: _____   Date: _____

STEM RESEARCH NOTEBOOK ENTRIES #5 AND #6, PAGE 1 (LESSON PLAN 1)

PHYSICS IN MOTION GAME DAYS

Circle your predictions before doing the activity. Then, circle your observations.

Questions	Predictions	Observations
1. What direction will the toy move when I give it a medium push?	Forward Backward Right Left	Forward Backward Right Left
2. How far will the toy move when I give it a medium push?	It will not reach the finish line. It will reach the finish line. It will move past the finish line.	It did not reach the finish line. It reached the finish line. It moved past the finish line.
3. How far will the toy move when I flick it with my finger?	It will not reach the finish line. It will reach the finish line. It will move past the finish line.	It did not reach the finish line. It reached the finish line. It moved past the finish line.

Name: _____ Date: _____

STEM RESEARCH NOTEBOOK ENTRIES #5 AND #6, PAGE 2 (LESSON PLAN 1)

PHYSICS IN MOTION GAME DAYS

Circle your predictions before doing the activity. Then, circle your observations.

Questions	Predictions	Observations
4. How far will the toy move when I blow on it?	It will not reach the finish line. It will reach the finish line. It will move past the finish line.	It did not reach the finish line. It reached the finish line. It moved past the finish line.
5. Will there be differences in movement between the bigger and smaller toys when I push them with a medium push?	Yes (circle one): The bigger toy will move farthest. The smaller toy will move farthest. No	Yes (circle one): The bigger toy moved farthest. The smaller toy moved farthest. No
6. Will the toys need a push to move down the ramp?	Yes No	Yes No
7. Will the toys move faster on the higher or lower ramp?	Higher Lower	Higher Lower

Name: _____   Date: _____

STEM RESEARCH NOTEBOOK ENTRIES #7 AND #8, PAGE 1 (LESSON PLAN 1)

PLAYGROUND PALS

Circle your predictions before doing the activity. Then, circle your observations.

Questions	Predictions	Observations
SWING Will you swing higher when you are pushed by a friend or when you pump your own legs?	Pushed by friend Pump own legs	Pushed by friend Pump own legs
SLIDE What can you put under you to make you slide the fastest?	Nothing Carpet Cardboard Waxed paper	Nothing Carpet Cardboard Waxed paper

Name: _____   Date: _____

STEM RESEARCH NOTEBOOK ENTRIES #7 AND #8, PAGE 2 (LESSON PLAN 1)

PLAYGROUND PALS

Circle your predictions before doing the activity. Then, circle your observations.

Questions	Predictions	Observations
SEESAW What happens when students of similar size sit on each end of the seesaw?	We will not move. One side will go up.	We did not move. One side went up.
SEESAW What happens when students of different sizes sit on each end of the seesaw?	We will not move. The side with the larger student will go up. The side with the smaller student will go up.	We did not move. The side with the larger student went up. The side with the smaller student went up.

Name: _____   Date: _____

STEM RESEARCH NOTEBOOK ENTRY #9 (LESSON PLAN 2)

Draw a picture of a roller coaster:

Name: _____   Date: _____

STEM RESEARCH NOTEBOOK ENTRY #10 (LESSON PLAN 2)

I learned ...

NATIONAL SCIENCE TEACHING ASSOCIATION

Name: _____

Date: _____

STEM RESEARCH NOTEBOOK ENTRY #11 (LESSON PLAN 2)

Our best roller coaster design looked like this:

Name: _____   Date: _____

STEM RESEARCH NOTEBOOK ENTRY #12 (LESSON PLAN 2)

I learned that ...

Gravity makes a roller coaster car move in this direction (circle one):

Friction makes a roller coaster car (circle one):   Go faster   Slow down

Draw a picture of a section of your roller coaster track where gravity was at work:

Draw a picture of a section of your roller coaster track where your car stopped:

Name: _____

Date: _____

STEM RESEARCH NOTEBOOK ENTRY #13 (LESSON PLAN 3)

PLANNING YOUR ROLLER COASTER

Draw a picture of a roller coaster design for your team:

Name: _____   Date: _____

STEM RESEARCH NOTEBOOK ENTRY #14 (LESSON PLAN 3)

Draw a picture of your team's final track design:

Then, answer these questions:

We know this track design was the best because

_____

_____

_____

_____

We saw gravity at work when

_____

_____

_____

APPENDIX B

OBSERVATION, STEM RESEARCH NOTEBOOK, AND PARTICIPATION RUBRIC

Name: _____

Observation, STEM Research Notebook, and Participation Rubric

Categories (components)	Missing or Unrelated (0 points)	Beginning (1 point)	Developing (2 points)	Meets Expectations (3 points)	Exceeds Expectations (4 points)	Score
OBSERVATION OF LISTENING AND DISCUSSION SKILLS	Component is missing or unrelated.	Does not listen to others and shows little respect for alternative viewpoints.	Occasionally listens to others but often speaks out of turn.	Listens to others, only occasionally speaks out of turn, and generally accepts other points of view.	Listens carefully to others, waits for turn to speak, and respects alternative viewpoints.	
STEM RESEARCH NOTEBOOK	Component is missing or unrelated.	Demonstrates little understanding of the concepts being taught.	Recalls and is able to explain basic facts and concepts.	Demonstrates ability to apply concepts, using information in new situations.	Demonstrates a deep understanding of concepts by drawing relationships between ideas and using information to generate new ideas.	
PARTICIPATION	Component is missing or unrelated.	Does not volunteer. When responding to teacher prompts, comments are sometimes not relevant to the discussion.	Responds to teacher prompts during classroom discussions but does not volunteer.	Willingly participates in classroom discussions and offers relevant comments.	Contributes insightful comments and poses thoughtful questions.	

TOTAL SCORE: _____

COMMENTS:

APPENDIX C

CONTENT STANDARDS ADDRESSED IN THIS MODULE

NEXT GENERATION SCIENCE STANDARDS

Table C1 (p. 112) lists the science and engineering practices, disciplinary core ideas, and crosscutting concepts this module addresses. The supported performance expectations are as follows:

- K-PS2-1. Plan and conduct an investigation to compare the effects of different strengths or different directions of pushes and pulls on the motion of an object.

- K-PS2-2. Analyze data to determine if a design solution works as intended to change the speed or direction of an object with a push or a pull.

Table C1. *Next Generation Science Standards (NGSS)*

Science and Engineering Practice
PLANNING AND CARRYING OUT INVESTIGATIONS Planning and carrying out investigations to answer questions or test solutions to problems in K–2 builds on prior experiences and progresses to simple investigations, based on fair tests, which provide data to support explanations or design solutions. • With guidance, plan and conduct an investigation in collaboration with peers.
Disciplinary Core Ideas
PS2.A. FORCES AND MOTION • Pushes and pulls can have different strengths and directions. • Pushing or pulling on an object can change the speed or direction of its motion and can start or stop it. **PS2.B. TYPES OF INTERACTIONS** • When objects touch or collide, they push on one another and can change motion. **PS3.C. RELATIONSHIP BETWEEN ENERGY AND FORCES** • A bigger push or pull makes things speed up or slow down more quickly.
Crosscutting Concept
CAUSE AND EFFECT • Simple tests can be designed to gather evidence to support or refute student ideas about causes.

Source: NGSS Lead States. 2013. *Next Generation Science Standards: For states, by states.* Washington, DC: National Academies Press. *www.nextgenscience.org.*

Table C2. Common Core Mathematics and English Language Arts (ELA) Standards

MATHEMATICAL PRACTICES	READING STANDARDS

MATHEMATICAL PRACTICES

- MP1. Make sense of problems and persevere in solving them.
- MP2. Reason abstractly and quantitatively.
- MP3. Construct viable arguments and critique the reasoning of others.
- MP4. Model with mathematics.
- MP5. Use appropriate tools strategically.
- MP6. Attend to precision.

MATHEMATICAL CONTENT

- K.MD.B.3. Classify objects into given categories; count the numbers of objects in each category and sort the categories by count.
- K.CC.C.6. Identify whether the number of objects in one group is greater than, less than, or equal to the number of objects in another group, e.g., by using matching and counting strategies.
- K.CC.C.7. Compare two numbers between 1 and 10 presented as written numerals.
- K.CC.B.4. Understand the relationship between numbers and quantities; connect counting to cardinality.
- K.CC.B.4a. When counting objects, say the number names in the standard order, pairing each object with one and only one number name and each number name with one and only one object.
- K.CC.B.4b. Understand that the last number name said tells the number of objects counted. The number of objects is the same regardless of their arrangement or the order in which they were counted.
- K.CC.B.4c. Understand that each successive number name refers to a quantity that is one larger.
- K.MD.A.1. Describe measurable attributes of objects, such as length or weight. Describe several measurable attributes of a single object.
- K.MD.A.2. Directly compare two objects with a measurable attribute in common, to see which object has "more of"/"less of" the attribute, and describe the difference. For example, directly compare the heights of two children and describe one child as taller/ shorter.

READING STANDARDS

- RI.K.1. With prompting and support, ask and answer questions about key details in a text.
- RI.K.3. With prompting and support, describe the connection between two individuals, events, ideas, or pieces of information in a text.

WRITING STANDARDS

- W.K.2. Use a combination of drawing, dictating, and writing to compose informative/explanatory texts in which they name what they are writing about and supply some information about the topic.
- W.K.5. With guidance and support from adults, respond to questions and suggestions from peers and add details to strengthen writing as needed.
- W.K.7. Participate in shared research and writing projects (e.g., explore a number of books by a favorite author and express opinions about them).

SPEAKING AND LISTENING STANDARDS

- SL.K.1. Participate in collaborative conversations with diverse partners about *kindergarten topics and texts* with peers and adults in small and larger groups.
- SL.K.3. Ask and answer questions in order to seek help, get information, or clarify something that is not understood.
- SL.K.5. Add drawings or other visual displays to descriptions as desired to provide additional detail.

Source: National Governors Association Center for Best Practices and Council of Chief State School Officers (NGAC and CCSSO). 2010. *Common core state standards.* Washington, DC: NGAC and CCSSO.

Table C3. National Association for the Education of Young Children (NAEYC) Standards

NAEYC Curriculum Content Area for Cognitive Development: Science and Technology
• 2.G.02. Children are provided varied opportunities and materials to learn key content and principles of science.
• 2.G.03. Children are provided with varied opportunities and materials that encourage them to use the five senses to observe, explore, and experiment with scientific phenomena.
• 2.G.04. Children are provided with varied opportunities to use simple tools to observe objects and scientific phenomena.
• 2.G.05. Children are provided with varied opportunities and materials to collect data and to represent and document their findings (e.g., through drawing or graphing).
• 2.G.06. Children are provided with varied opportunities and materials that encourage them to think, question, and reason about observed and inferred phenomena.
• 2.G.07. Children are provided with varied opportunities and materials that encourage them to discuss scientific concepts in everyday conversation.
• 2.G.08. Children are provided with varied opportunities and materials that help them learn and use scientific terminology and vocabulary associated with the content areas.
• 2.H.03. Technology is used to extend learning within the classroom and integrate and enrich the curriculum.

Source: National Association for the Education of Young Children (NAEYC). 2005. *NAEYC early childhood program standards and accreditation criteria: The mark of quality in early childhood education.* Washington, DC: NAEYC.

Table C4. 21st Century Skills From the Framework for 21st Century Learning

21st Century Skills	Learning Skills and Technology Tools	Teaching Strategies	Evidence of Success
INTERDISCIPLINARY THEMES	• Economic, Business, and Entrepreneurial Literacy • Health Literacy • Environmental Literacy	• Facilitate student use of the engineering design process (EDP) to plan, build, and maintain marble track roller coasters. • Discuss the importance of safety guidelines for roller coasters and facilitate students' development of safety rules for their own roller coaster designs. • Have students observe changes that occur in the environment with the construction of amusement parks.	• Students demonstrate an understanding of the EDP and use it successfully to create marble track roller coasters. • Students create safety rules for their own roller coasters.
LEARNING AND INNOVATION SKILLS	• Creativity and Innovation • Critical Thinking and Problem Solving • Communication and Collaboration	• Introduce the EDP as a problem-solving framework. • Facilitate critical thinking and problem-solving skills through having students build and improve marble track roller coasters.	• Students demonstrate an understanding of the EDP through teamwork to create marble track roller coasters. • Students demonstrate creativity and innovation, critical thinking and problem solving, communication, and collaboration as they plan, build, test, and improve marble track roller coasters.
INFORMATION, MEDIA, AND TECHNOLOGY SKILLS	• Information Literacy • Media Literacy • Information, Communications, and Technology Literacy	• Engage students in guided practice and scaffolding strategies through the use of developmentally appropriate books, videos, and websites to advance their knowledge.	• Students acquire and use deeper content knowledge via information, media and technology skills as they create marble track roller coasters.

Continued

Table C4. (*continued*)

21st Century Skills	Learning Skills and Technology Tools	Teaching Strategies	Evidence of Success
LIFE AND CAREER SKILLS	• Flexibility and Adaptability • Initiative and Self-Direction • Social and Cross-Cultural Skills • Productivity and Accountability • Leadership and Responsibility	• Facilitate student collaborative group work to foster life and career skills.	• Throughout this module, students collaborate to conduct investigations and create marble track roller coasters.

Source: Partnership for 21st Century Learning, Battelle for Kids. 2015. Framework for 21st Century Learning. *www.battelleforkids.org/networks/p21/frameworks-resources.*

Table C5. English Language Development (ELD) Standards

ELD STANDARD 1: SOCIAL AND INSTRUCTIONAL LANGUAGE English language learners communicate for Social and Instructional purposes within the school setting. **ELD STANDARD 2: THE LANGUAGE OF LANGUAGE ARTS** English language learners communicate information, ideas and concepts necessary for academic success in the content area of Language Arts. **ELD STANDARD 3: THE LANGUAGE OF MATHEMATICS** English language learners communicate information, ideas and concepts necessary for academic success in the content area of Mathematics. **ELD STANDARD 4: THE LANGUAGE OF SCIENCE** English language learners communicate information, ideas and concepts necessary for academic success in the content area of Science. **ELD STANDARD 5: THE LANGUAGE OF SOCIAL STUDIES** English language learners communicate information, ideas and concepts necessary for academic success in the content area of Social Studies.

Source: WIDA. 2012. 2012 amplification of the English language development standards: Kindergarten–grade 12. *https://wida.wisc.edu/teach/standards/eld.*

INDEX

Page numbers printed in **boldface type** indicate tables, figures, or handouts.